STUDENT'S SOLUTIONS MANUAL TO ACCOMPANY

MULTIPLE-CHOICE AND FREE-RESPONSE QUESTIONS IN PREPARATION FOR THE AP STATISTICS EXAMINATION

(SECOND EDITION)

By

Michael Allwood
Brunswick School
Greenwich, CT

Bonnie Montgomery
Morris Knolls High School
Denville, NJ

D&S MARKETING SYSTEMS, INC.
1205 38th Street Brooklyn, NY 11218
www.dsmarketing.com

ISBN # 1-934780-03-0

Printed in the U.S.A.

PREFACE

This Student's Solutions Manual, a supplement to MULTIPLE-CHOICE & FREE-RESPONSE QUESTIONS IN PREPARATION FOR THE AP STATISTICS EXAMINATION, provides students and teachers with detailed solutions to the questions in the five sample examinations. For the multiple-choice questions, in addition to an explanation of the correct solution, an indication is often provided as to why the other options are incorrect. Throughout the manual, where applicable, instructions are included as to how to use a graphing calculator as an aid to the solution of the problems.

The authors feel certain that students will benefit greatly from taking the practice exams in the question book, and from use of this solutions manual for checking their answers.

Our thanks go to our editors, Mary Mortlock and Daren Starnes, and to Tammi Pruzansky of D&S Marketing Systems, Inc., who typeset the book and the manual, and prepared the materials for printing.

All communications concerning this book should be addressed to:

D&S Marketing Systems, Inc.
1205 38th Street
Brooklyn, NY 11218
www.dsmarketing.com

TABLE OF CONTENTS

Sample Examination One – Answers

1. **(D)** The upper half of the boxplot is much longer than the lower half, and so the distribution is positively skewed. Referring to the boxplot again, the IQR is $Q_3 - Q_1 \approx 2.7 - 0.7 = 2$. So the IQR of the pay raises is approximately \$2000.

2. **(A)** This is cluster sampling, with each volume forming a cluster. (To avoid confusion, note that the word "volume" here refers to one of the 30 books that make up the encyclopedia.) Do not get cluster sampling confused with stratified random sampling. If the researcher believed that the various volumes were *different* in terms of the proportion of diagrams and pictures, then she might take a random sample of pages from *every volume*. That would be stratified random sampling.

3. **(A)** The null hypothesis must reflect the company's claim that the mean breaking strength is, in fact, 80 pounds. Since the group wants to find out whether there is evidence that the cloth's breaking strength is *less* than 80 pounds, the alternative hypothesis is as given in (A).

4. **(D)** Suppose that an employee is chosen at random. Let M be the event that the employee opts for medical insurance and let L be the event that the employee opts for life insurance.

   We are given that $P(M) = 0.78, P(L) = 0.42,$ and $P(M \cup L) = 0.82$. The "addition rule" states that $P(A \cup B) = P(A) + P(B) - P(A \cap B)$, or, solving for $P(A \cap B)$, that $P(A \cap B) = P(A) + P(B) - P(A \cup B)$. So, applying that rule,
   $P(M \cap N) = P(M) + P(N) - P(M \cup N) = 0.78 + 0.42 - 0.82 = 0.38.$

5. **(E)** By drawing a vertical line up from 43 on the horizontal axis to the curve and then drawing a horizontal line from this point of intersection to the vertical axis, we see that the percentage of members *less than* or equal to 43 years old is approximately 25%, that is, approximately 40 members. So the number of members *more than* 43 years old is approximately 120.

6. **(B)** Two independent random samples are being taken and their means are being compared. So the two-sample t-test for means would be used.

   Let us now look at why the other options are incorrect. Considering (D) and (E), note that the coach wishes to compare the thrower's average (mean) length of throw in the morning with his/her average (mean) length of throw in the afternoon. The tests in (D) and (E) are for proportions, and are therefore not applicable to this situation. The test in (A) is a *one*-sample test which therefore cannot be used to make this comparison. Lastly, looking at (C), the paired t-test is used when you have a single sample with two data values for each element in the sample or when you have an experiment that uses matched pairs. Neither of these scenarios is present in this context, and so option (C) is incorrect.

7. **(B)** If, for example, she sets 6 places, then the probability that all the children can be seated is the probability that 6 or fewer children show up. This is the sum of the probabilities up to and including 6, which is 0.78, and this is not quite enough. Diana wants to be 80% sure, so she will have to set 7 places.

8. **(A)** This is a one proportion z-test. The null hypothesis is $H_0: p = 0.1$, and the z-statistic is calculated assuming H_0 to be true. So the z-statistic is

   $z = \dfrac{\hat{p} - p}{\sqrt{\dfrac{p(1-p)}{n}}} = \dfrac{0.08 - 0.1}{\sqrt{\dfrac{(0.1)(0.9)}{200}}}$. We're doing a two-tailed test since we're asking whether the proportion of defective springs has *changed*, and the p-value is twice the area in the *left* tail of the normal distribution, since the sample proportion is *less* than the hypothesized population proportion. Hence, the p-value is $2 \cdot P\left(z < \dfrac{0.08 - 0.1}{\sqrt{\dfrac{(0.1)(0.9)}{200}}}\right)$.

   (The conditions for use of the one-proportion z-test are satisfied since we are given that the sample is random, and $np = 200(0.1) = 20 \geq 10$, $n(1-p) = 200(0.9) = 180 \geq 10$.

   Note: If we were asked for a *confidence interval* (with, let's say, a 95% confidence level), then the formula would be $0.08 \pm 1.96\sqrt{\dfrac{(0.08)(0.92)}{200}}$. The square root term used in the hypothesis test is different from the one in the confidence interval.

9. **(A)** P has an extremely large x-coordinate relative to those of the other data points. Removal of a point with an extreme x-coordinate will have a large effect on the regression line if the point is not close to the regression line that would pertain to the remaining points. This is clearly the case here. What is more, it is obvious that without P the regression line would have a greater slope than the one shown in the diagram. However, Q has an x-coordinate that is near the mean x-coordinate for all the data points and has an extreme y-coordinate. So the effect of the removal of Q would be to lower the regression line, but only very slightly as there are so many other points.

10. **(C)** This is a chi-square test for goodness of fit.

	Blue	Brown	Green
Expected Count	16	14	10
Observed Count	10	19	11

So we have $\chi^2 = \sum \frac{(\text{observed} - \text{expected})^2}{\text{expected}} = \frac{(10-16)^2}{16} + \frac{(19-14)^2}{14}$ $\frac{(11-10)^2}{10} = 4.136$. There are $3 - 1 = 2$ degrees of freedom. So, using χ^2cdf(4.136,999,2), or the equivalent, we get a p-value of 0.126. (On the TI-83 and 84 calculators, "χ^2cdf(" is available in the menu accessed by selecting 2nd,DISTR.)

11. **(E)** The number of trials, X, up to and including the first success has a geometric distribution.
$P(X \geq 3) = 1 - P(X = 1) - P(X = 2) = 1 - (0.2) - (0.8)(0.2) = 0.64.$
Alternatively, taking three or more trials to get the first success is equivalent to failing on the first two trials. The probability that this happens is $(0.8)(0.8) = 0.64$.

12. **(B)** A small p-value indicates strong evidence against H_0 and in favor of H_a. We have no evidence at all to support the alternative hypotheses in (A) and (E). With a sample mean of 10.8 we clearly have stronger evidence for $\mu > 10$ than for $\mu < 11$; so the p-value for (B) is smaller than the p-value for (C). The p-value for (D) is twice the p-value for (C). So the p-value for (B) is smallest.

13. **(D)** The mean, the standard deviation, and the range will all be affected by the outlier. Thus, options (A), (B), (C), and (E) are all incorrect, and so the correct response is (D).

14. **(C)** The number of underclassmen who take the bus to school is $638 \cdot (0.831) = 530$ and the number of upperclassmen who take the bus to school is $523 \cdot (0.709) = 371$. So the required probability is $530/(530 + 371) = 0.588$.

15. **(C)** In any hypothesis test, the p-value is the probability of getting a value of the test statistic at least as extreme (in favor of H_a) as the one obtained, given that H_0 is true. Here H_0 states that $p_M = p_W$, and so the correct response is (C).

16. **(B)** We are told that the Board wishes to determine the opinions of the parents of students in grades K-12 on the issue. Therefore, sending the message to parents of students who are not in the 12th grade is not a source of bias, and so (B) is the correct response.

    The statement in (A) is correct (and therefore (A) is an incorrect response) since the email message includes the sentence, "The reduction in parking would allow for a substantial expansion in student activities." This comment is likely to influence the reader in favor of not allowing 12th graders to drive to school. Likewise, the issues sited in (C), (D), and (E) are clearly all potential sources of bias, and so these options are incorrect.

17. **(C)** The correlation is unaffected by adding a constant to all the x-values (or y-values) or by multiplying all the x-values (or y-values) by a constant. For any given student, let x be the height and y be the weight. When the heights are converted to centimeters and the weights are converted to kilograms, the x-values are multiplied by 2.54 and the y-values are multiplied by 0.454. So the value of the correlation is unchanged.

18. **(E)** Saying that the distribution of A's throws is positively skewed does not tell us anything about how far A is throwing. It merely tells us that the spread of the throws above the median is greater than the spread of the throws below the median. Thus, the correct response is (E).

    Let us look now at why the statements in the other options are correct (making these incorrect responses). Clearly the scenarios in (A) and (B) would be valid reasons for choosing thrower A over thrower B. Considering (D), since a javelin competition is judged according to a thrower's maximum throw, the maximum for thrower A being greater than the maximum for thrower B is a good reason for choosing thrower A over thrower B. Looking at (C), suppose, for example, that the upper quartiles for throwers A and B were 45 meters and 40 meters respectively. This would mean that 25% of the time thrower A is throwing the javelin at least 45 meters, while 25% of the time thrower B is throwing at least 40 meters, which is clearly less good than A's achievement.

19. **(B)** The two explanatory variables are the drug administered and the number of doses.

20. **(C)** Note that the confidence interval tells us something about the population *mean*. The statement in (C), however, tries to say something about the actual flight times, without reference to the population mean. This is therefore a false statement, and thus (C) is the correct response.

    Options (A), (B), (D), and (E) are all correct statements regarding the population mean (making these responses incorrect).

21. **(B)** Both statistics (the sample mean and the sample median) are unbiased since their expected values (means) are both 3, which is the value of the parameter that is being estimated. We prefer the statistic that has the smaller standard deviation, since this tells us that this statistic is more likely than the other statistic to be *close* to μ.

22. **(B)** To say that two events are mutually exclusive means that they can't both happen. In this case, A and B *can* both happen (the card can be the queen of hearts) so A and B are not mutually exclusive. To say that two events A and B are independent is to say that knowing that one of them is happening does not change the probability that the other is happening; in other words that $P(A \mid B) = P(A)$, or, equivalently, $P(B \mid A) = P(B)$. Here, the probability that the card is a queen is 4/52 = 1/13, and the probability that the card is a queen given that it is a heart is also 1/13. So the events are independent.

23. **(D)** We know that the study is conducted in at least a *single* blind manner as we are told that the people who measure the muscle mass do not know which subjects are in which group. However, it can't be *double* blind as the subjects clearly know whether or not they are doing the warm-up exercises.

    The statement in (A) is correct (making (A) an incorrect response), since the treatments (doing or not doing the warm-up exercises) are imposed on the volunteers. The statement in (B) is clearly correct (making (B) an incorrect response) since the volunteers are randomly assigned to the groups. The statement in (C) is correct (making (C) an incorrect response) since one of the groups does no warm-up exercises, thus making this group a control group. The statement in (E) is correct (making (E) an incorrect response) since there is indeed no blocking in the description of the experiment.

24. **(B)** If $y = ax + b$, where a and b are constants and $a > 0$, then $s_y = a \cdot s_x$. So here $s_y = 0.7s_x = 0.7 \cdot 8.8 = 6.16$.

    Alternatively, multiplying the x-values by 0.7 has the effect of multiplying the standard deviation by 0.7, and adding 30 has no effect on the standard deviation. So the new standard deviation is $0.7 \cdot 8.8 = 6.16$.

25. **(C)** r^2 is the proportion of the y-variability that can be explained by the regression line. (Here's an explanation of this in the context of this question: If the math scores and the physics scores are linearly related, then as we think of the math scores varying we expect the physics scores to vary accordingly. However, the actual physics scores do not vary exactly according to the approximate linear relationship: some are above the line, some below. So some of the variation in physics scores can be explained by their approximate linear relationship to the varying math scores, some cannot. The proportion of the variation of the physics scores that *can* be explained by the variation of the math scores is r^2.)

    Let's look now at the other options. Option (A) is incorrect as a high value of r^2 could indicate a large *negative* correlation. Option (B) is incorrect because a high value of r^2 (or, equivalently, a value of r that is close to 1 or -1) does not mean that a straight line is the *best* model – you might still find a curve that fits the data better. The value of r^2 can give no information about the presence (or not) of outliers, and so option (D) is incorrect. Finally, considering option (E), a comparison between the physics scores and the math scores would be achieved by looking at the pattern of points in the scatterplot relative to the line $y = x$. Assuming that the math scores are denoted by x and the physics scores by y, if the points are generally above the line $y = x$ then, generally speaking, the physics scores are higher than the math scores. If the points are generally below the line $y = x$, then, generally speaking, the physics scores are lower than the math scores. However, the value of r^2 gives no information about a comparison between the scores in the two subjects since, for example, a value of r^2 close to 1 could indicate a strong linear relationship between the two variables with the points generally above the line $y = x$ or with the points generally below the line $y = x$.

26. **(D)** The column total for "Democrat" is 182, the row total for "Female" is 210, and the grand total is 400. So the expected number of female Democrats is $182 \cdot 210/400 = 95.55$.

27. **(D)** In order to show why the result does not imply that in order to live longer one should start to regularly consume substantial amounts of olive oil, we have to explain how olive oil consumption can be *associated* with long life without olive oil consumption actually *causing* long life. The argument in (D) is the only one that does this. Note that option (C) is not adequate, because it contains no suggestion that regular exercise (or any other life prolonging habit) is *connected with olive oil consumption*, and therefore does not explain how olive oil consumption can be associated with long life without causing it.

28. **(E)** A scatterplot deals with the situation where you have one set of objects (or people) and *two* measurements for each object. For example, a scatterplot could illustrate, for one set of children, the connection between their on-task times and their spelling scores. However, here we have two sets of children being compared, and so a scatterplot would not be suitable; thus, (E) is the correct response. (All of the graphs given in the other four options would be suitable for making this comparison.)

29. **(D)** Let the amount of flour per bag have mean μ. We are told that the distribution has standard deviation $\sigma = 0.4$. We estimate μ using the sample mean, $\bar{x}$. If the sample size is large (we hope to get an answer bigger than 30), then the distribution of $\bar{x}$ is approximately normal with mean μ and standard deviation $\sigma/\sqrt{n}$. For any normally distributed random variable, with 95% probability the random variable will be within 1.96 standard deviations of the mean. (We get 1.96 using invNorm(.975), or the equivalent; on the TI-83 and 84 calculators, "normalcdf(" is accessed by selecting 2nd,DISTR,2.) So, with 95% probability, $\bar{x}$ will be within $1.96 \cdot \sigma/\sqrt{n}$ of μ. So, referring to the question, we need

$$1.96 \cdot \frac{\sigma}{\sqrt{n}} = 0.1,$$

that is,

$$1.96 \cdot \frac{0.4}{\sqrt{n}} = 0.1.$$

So

$$\sqrt{n} = \frac{1.96 \cdot 0.4}{0.1},$$

giving

$$n = \left(\frac{1.96 \cdot 0.4}{0.1}\right)^2 = 61.47.$$

So we need $n = 62$.

30. **(E)** All three statements are true. Note that the question states that we are talking about small samples – if the sample were large we would not need to make the assumptions in I and II. With regard to II, it's important to remember that the *t*-test does indeed require the assumption that the population distribution is normal.

31. **(D)** The number of seeds that germinate is binomially distributed with $n = 20$ and $p = 0.7$. The mean of this binomial distribution is given by $np = 20 \cdot 0.7 = 14$ and the standard deviation is given by $\sqrt{np(1-p)} = \sqrt{20 \cdot 0.7 \cdot 0.3} = 2.049$.

    Be careful not to confuse the number of seeds that germinate with the *proportion* of seeds that germinate. The values given in (A) are the correct mean and standard deviation for the *proportion* of seeds that germinate.

32. **(D)**

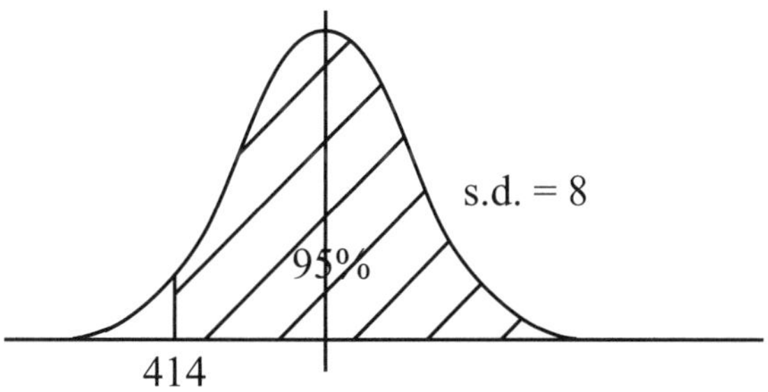

We are given that 95% of the cans contain more than 414 ml, which means that 5% contain less than 414 ml. So, using invNorm(.05), or the equivalent, we see that at 414, $z = -1.645$. (On the TI-83 and 84 calculators, "invNorm(" is accessed by selecting 2nd,DISTR,3.) Thus, using $z = (x - \mu)/\sigma$, we see that

$$-1.645 = \frac{414 - \mu}{8}.$$

Solving for μ, we get $\mu = 414 + 1.645 \cdot 8 = 427.16$. Option (D) gives the closest approximation to this answer.

33. **(D)** The confidence interval is given by $\hat{p} \pm z * \sqrt{\frac{\hat{p}(1-\hat{p})}{n}}$, and the margin of error is $z * \sqrt{\frac{\hat{p}(1-\hat{p})}{n}}$. So in order to make the margin of error small, we need to minimize $z*$ and maximize n. The larger value of n available to us is 500, and $z*$ will be minimized by taking the smallest available confidence level – so we choose 95%.

34. **(C)** This question is about blocking. The belief is that the younger a person is, the more likely he is to respond to a drug of this sort – so it makes sense to block by age. This is done by forming a block consisting of the three young men, a block consisting of the three middle-aged men, and a block consisting of the three elderly men. The three treatments are then randomly assigned within each block. The reason this is helpful is that we are now giving drug A to one young man, one middle aged man, and one elderly man; and the same for each of the other two drugs. So, in terms of age, the treatment groups are very similar. Therefore, if, for example, the men treated with drug A do better than the others, we can more readily attribute this to the greater effectiveness of drug A, and not to differences in age.

    Another way of looking at this, is that in each of the procedures described in (A), (B), (D), and (E), there is some possibility that the treatment groups will differ from each other in terms of age. Then, if, for example, one treatment group responds significantly better, on average, than another treatment group, it would be impossible to tell whether the difference comes about as a result of the greater effectiveness of one of the drugs, or of the age differences between the groups.

35. **(A)** We are given that the *p*-value is 0.063, which is not particularly small (this being judged by the fact that 0.063 is greater than the significance level, 0.05). This means that if the population mean were 40,000, then it would not be particularly unlikely that the sample mean would be as far away from 40,000 as the one obtained in the study. Thus, we do not have sufficient evidence to conclude that the mean life of the tires is not equal to 40,000, and option (A) is correct. Note that we never "accept H_0," so option (C) is incorrect. (In this context, accepting H_0 would mean that we have sufficient evidence to conclude that the mean life of the tires was actually equal to 40,000 miles *exactly*. This is very, very unlikely to be the case.) Turning our attention to (B), the alternative hypothesis in the question tells us that a two-tailed test is being used. The conclusion in (B) is the conclusion for the equivalent *one*-tailed test, and is therefore incorrect.

36. **(B)** The quantity $\sigma/\sqrt{n}$ is the standard deviation of the sampling distribution of the sample mean.

37. **(B)** Firstly, increasing the significance level, α, increases the probability of a Type I error and decreases the probability of a Type II error. So we choose $\alpha = 0.05$. Secondly, the probability of a Type II error is reduced by increasing the sample size, so we choose $n = 40$. Thirdly, looking at the true value of μ, we want to reduce the probability of a Type II error, so we want to increase the probability of correctly rejecting H_0 in favor of H_a, which says $\mu < 50$. The further the true value of μ is to the left of 50, the more likely we are to reject H_0. So we select $\mu = 46$.

38. **(E)** The confidence interval for the slope is $b \pm t^* \cdot s_b$. First, b is the slope of the regression line, and its value is given by the table to be 0.264. Second, t^* is the critical value of t for a 99% confidence level with $n - 2 = 22$ degrees of freedom, which the t-table provided with the exam gives to be 2.819. Third, s_b is the standard error of the slope of the regression line, whose value is given in the table as 0.109.

39. **(E)** Since the sample size is large ($n = 40$), the Central Limit Theorem tells us that the distribution of the sample mean, $\bar{x}$, is approximately normal. The mean of the sampling distribution of $\bar{x}$ is $\mu = 4$ and the standard deviation is $\sigma/\sqrt{n} = 2.309/\sqrt{40} = 0.36508$. We need $P(\bar{x} < 5)$.

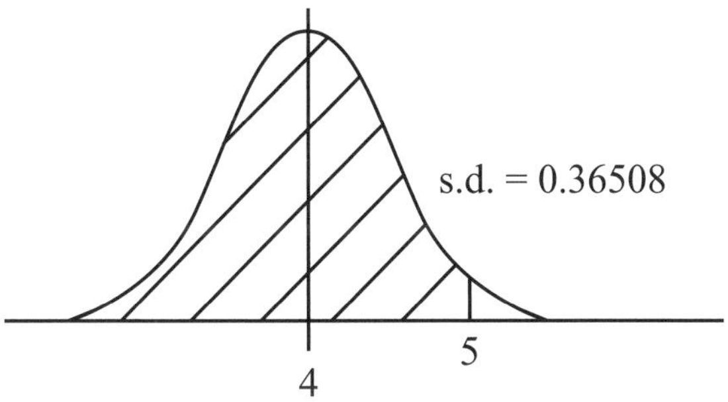

For $\bar{x} = 5, z = (5 - 4)/0.36508 = 2.739$. So, using normalcdf(-999,2.739), or the equivalent, we find that $P(\bar{x} < 5) = 0.997$. (On the TI-83 and 84 calculators, "normalcdf(" is accessed by selecting 2nd,DISTR,2.)

40. **(A)** The random pattern in the residual plot obtained by plotting log y against log x indicates a linear relationship between log y and log x. At this point, we can either use memory of the fact that this linear relationship indicates a power relationship between y and x, that is $y = ax^p$, or we can use the following mathematical argument:

    The linear relationship between log y and log x can be written in the form

$$\log y = A + B \log x.$$

Using the definition of logs:	$y = 10^{A + B\log x}$
Using $x^{a+b} = x^a \cdot x^b$:	$y = 10^A \cdot 10^{B\log x}$
Using $B\log x = \log x^B$:	$y = 10^A \cdot 10^{\log x^B}$
Using $10^{\log x^B} = x^B$:	$y = 10^A \cdot x^B$

    Writing 10^A as a and B as p, we have the relationship given in (A).

Section II

Question One

(a) Outliers are values that are more than 1.5 · (IQR) above Q_3 or more than 1.5 · (IQR) below Q_1. Here, IQR = $Q_3 - Q_1 = 667.5 - 463 = 204.5$. So

$Q_3 + 1.5 \cdot (\text{IQR}) = 667.5 + 1.5 \cdot (204.5) = 974.25$. The maximum file size is given to be 774, and 774 is less than 974.25, therefore there is no outlier at the upper end of the distribution. However, $Q_1 - 1.5 \cdot (\text{IQR}) = 463 - 1.5 \cdot (204.5) = 156.25$. The minimum file size is given to be 99 MB, which is smaller than 156.25, so there is at least one outlier at the lower end of the distribution.

[*There are other methods of identifying outliers that are also acceptable.*]

(b) The distribution is negatively skewed (skewed to the left).

(c) The table tells us that 463 MB is the first (lower) quartile file size and that 667.5 is the third (upper) quartile file size. Therefore, approximately 25% of the files have sizes less than 463 MB and approximately 75% of the files have sizes less than 667.5 MB. From this we can deduce that approximately 50% of the files have sizes between 463 MB and 667.5 MB.

(d) 167.26 MB is a typical deviation of the file sizes from the mean file size. Generally speaking, the standard deviation is only used for distributions that are roughly symmetrical and have no outliers. This distribution is not symmetrical and has outliers, so on two counts the standard deviation is not an appropriate measure of variability for this distribution.

[*Did you remember to put your interpretation of the standard deviation in context? An interpretation without the words "file size" would not get full credit!*]

Question Two

(a) $r = \pm\sqrt{0.82} = \pm\, 0.906$. Since the prices are declining as the age increases (or, alternatively, since the slope of the regression line is negative), the correlation is negative, and so $r = \boxed{-0.906}$. There is a strong negative linear relationship between the age and price of this type of car.

[*Did you include strength, direction, linearity, and context?*]

(b) The equation of the regression line is $\hat{y} = 25844.789 - 4764.155x$, where x is the age of the car and $\hat{y}$ is the predicted price.

[*Definition of x and $\hat{y}$ is required.*]

The slope is -4764.155, which tells us that when the age of the car increases by one year, the predicted price decreases by \$4764.155.

[*Note the word "predicted." Words like "average," "expected," or "estimated" would also be acceptable.*]

(c) The intercept is 25844.789. The predicted price when the age is zero is \$25844.789. Zero is outside the range of x-values on which the regression line is based, so there is no reason to think that the regression line should apply when the age is zero. So this value (\$25844.789) cannot be usefully applied to the prices of this type of car.

(d) The residual plot shows a random scatter, therefore a straight line is appropriate for modeling the prices of cars of this type that are at least one year and at most three years old.

Question Three

(a) The two funds have the same expected value of the monthly gain, but Fund A has the lower standard deviation, and so is likely to be less erratic – a more reliable form of income than Fund B.

[*Alternatively you could say that since Fund A has the smaller standard deviation, this fund is less likely to make a loss. Whatever your approach, you* do *need to mention that in terms of their means, the two funds are equal.*]

(b) We need the gain for Fund A to be greater than zero.

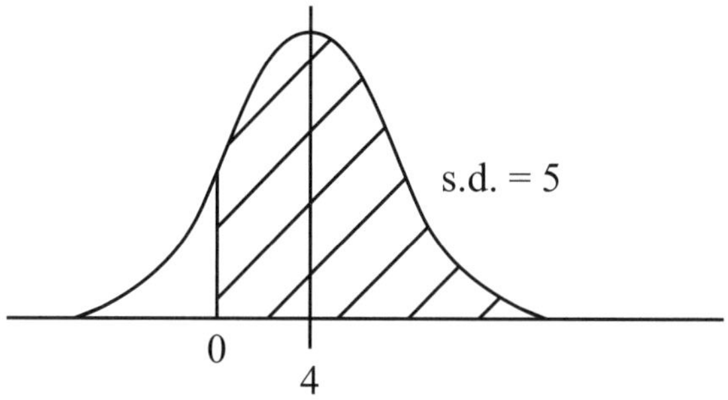

For a zero gain, the z-value is given by $z = (0 - 4)/5 = -0.8$. So, using normalcdf(-.8,999), or the equivalent, we find that the probability of Fund A gaining money is $\boxed{0.788}$. (On the TI-83 and 84 calculators, "normalcdf(" is accessed by selecting 2nd,DISTR,2. Alternatively, you could obtain the answer to this question by using normalcdf(0,999,4,5). Be sure, however, to show sufficient work, and do not use calculator syntax in your explanation.)

(c) The number of months for which Fund A gains money during the next four months is binomially distributed with $n = 4$ and $p = 0.78814$. So the required probability is ${}_4C_2 \cdot (0.78814)^2(0.21186)^2 = \boxed{0.167}$

(On the TI-83 and 84 calculators, ${}_4C_2$ is obtained by typing "4nCr2"; "nCr" is accessed by selecting MATH,PRB,3. Alternatively, the answer to this question could be obtained using binompdf(4,.78814,2), or the equivalent; "binompdf(" is available in the menu accessed by selecting 2nd,DISTR. Again, be sure to show your work, and do so without using calculator syntax; a correct answer with insufficient work would get no credit.)

(d) Let the gain in the first month on a \$1000 investment in Fund A be X, and let the equivalent gain for Fund B be Y. Then the gain for this investor is $Z = 8X + 3Y$.

So $\mu_Z = 8\mu_X + 3\mu_Y = 8 \cdot 4 + 3 \cdot 4 = \boxed{\$44}$

and $\sigma_Z = \sqrt{8^2\sigma_X^2 + 3^2\sigma_Y^2} = \sqrt{8^2 \cdot 5^2 + 3^2 \cdot 18^2} = \boxed{\$67.20}$.

Question Four

(a) Print the names of all 50 gymnasts on pieces of paper and place them in a hat. Then pick at random 25 names from the hat. These 25 gymnasts will form the group that receives the new treatment, and the remainder will be in the group that receives the current treatment.

[*Be sure to specify* which *group the first 25 names will go into. Very often students will say they go into "one group" without saying which group this is. Also, note that the method where you flip a coin for each gymnast and stop when you have 25 in one of the groups is not completely random assignment. For further details, please see "Know how to achieve random assignment" in Part II, Section 3 of "Top Tips for AP Statistics" in the Question Book.*]

(b) If some of the gymnasts were allowed to choose which group they would go into then it is quite possible, for example, that the more enthusiastic gymnasts who practice more would be keen to try out the new treatment. So then any greater toughness in the new treatment might be masked by the fact that those gloves were used more heavily. Random assignment produces two groups of subjects that are expected to be similar in all respects.

[*Be careful* not *to suggest that random assignment might* eliminate *differences between the groups.*]

(c) Despite the random assignment, in the first design there can be large differences between the individuals in the two groups and therefore the heaviness of use could differ greatly between the two treatments. The effect on the results of any difference in the toughness of the two treatments cannot be distinguished from the effect of the differences in the heaviness of use, and therefore any difference in toughness is harder to detect. However, under the second design, the two treatments will be subjected to almost identical use, and so any differences in the wear on the two types of glove can be attributed almost exclusively to the differences in the effectiveness of the two treatments. (The random assignment of the gloves to the hands means that any factors resulting from use by dominant or non-dominant hands are expected to be equally distributed between the two treatments.)

[*Remember – when comparing two things you must consider* both *in what you write.*]

Question Five

Hypotheses:
Let μ_1 and μ_2 be the population mean times when other participants are not present and present, respectively.
H_0: $\mu_1 = \mu_2$
H_a: $\mu_1 < \mu_2$

Check of conditions and naming of type of test:
This is a two-sample *t*-test.

The conditions for this two-sample *t*-test are:
1. Random assignment of subjects to treatments
2. Normally distributed populations or large treatment groups

We are told that the groups were assigned randomly. We do not have large treatment groups and so we will consider boxplots of the values for the groups.

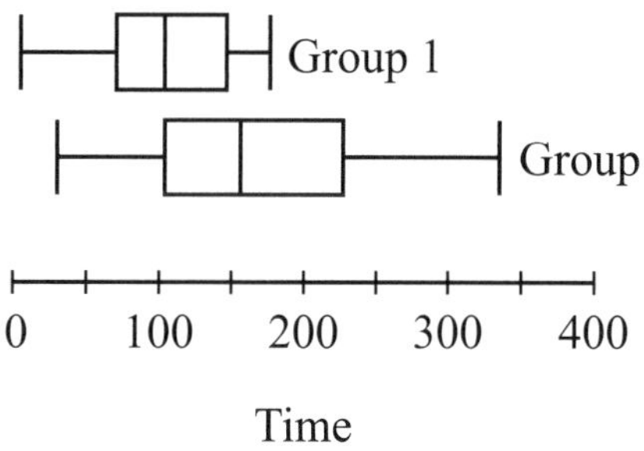

Since the distributions of the group times are roughly symmetrical and there are no outliers in either distribution, it appears reasonable to assume that the populations are normally distributed.

Mechanics:

$$t = \frac{\overline{X}_1 - \overline{X}_2}{\sqrt{\frac{s_1^2}{n_1} + \frac{s_2^2}{n_2}}} = \frac{145.5 - 220.063}{\sqrt{\frac{51.419^2}{16} + \frac{89.925^2}{16}}} = -2.879.$$

We use the *t* distribution with 23.861 degrees of freedom (given by the calculator). So the *p*-value is $P(t_{23.861} < -2.879) = 0.004$. (This is calculated using tcdf(-999,-2.879,23.861), or the equivalent; on the TI-83 and 84 calculators, "tcdf(" is accessed by selecting 2nd,DISTR,5.)

Conclusion:
Since the *p*-value is 0.004, which is less than 0.05, we have statistically significant evidence that the presence of other participants brings about a higher mean time for this task.

Question Six

(a) Hypotheses:

Let p be the proportion of students at Central HS who are in favor of the idea.

$H_0: p = 0.5$

$H_a: p > 0.5$

Check of conditions and naming of test:

This is a one-sample z-test for a proportion.

The conditions for this test are:

1. Simple random sample
2. np_0 and $n(1 - p_0)$ both greater than 10.

We are told that the sample of students is random. Moreover, $np_0 = 40 \cdot 0.5 = 20$ and $n(1 - p_0) = 40 \cdot 0.5 = 20$, which are both greater than 10.

Mechanics:

$$z = \frac{\hat{p} - p_0}{\sqrt{\dfrac{p_0(1 - p_0)}{n}}} = \frac{26/40 - 0.5}{\sqrt{\dfrac{(0.5)(0.5)}{40}}} = 1.897.$$

So the p-value is $P(Z > 1.897) = 0.029$. (This is calculated using normalcdf(1.897,999), or the equivalent; on the TI-83 and 84 calculators, "normalcdf(" is accessed by selecting 2nd,DISTR,2.)

Conclusion:

Since the p-value is 0.029 which is less than 0.05, we have statistically significant evidence that a majority of the students at Central HS are in favor of the idea.

(b) 1-3 will denote "D" (0 point), 4-6 will denote "N" (1 point), and 7-9 will denote "A" (2 points); ignore 0's.

[*Any assignment of digits that gives the correct probabilities is acceptable.*]

(c) By my assignment, and starting in the top left corner of the table and moving to the right, the responses are:

N (1) N (1) A (2) D (0) A (2) A (2) N (1) D (0) N (1) D (0) A (2) A (2) N (1) A (2) A (2) N (1) D (0) A (2) N (1) N (1),

giving a total score of 24.

(d) The 200 runs of the simulation give us an approximate probability for each of the total scores, given that the three responses are favored by equal proportions of the school. So, for example, $P(\text{total score} = 30) \approx 1/200 = 0.005$. Therefore, considering the results of the 200 runs of the simulation, if the three responses are favored by equal proportions of the school, the probability of obtaining a total score of at least 24 in a sample of 20 students is approximately $30/200 = 0.15$, which is not particularly small. It is therefore plausible, given that the three responses are favored by equal proportions of the school, to get a sample of 20 students with a total score of 24. Thus we do not have convincing evidence that the student body as a whole is in favor of the idea.

Sample Examination Two – Answers

1. **(D)** It is an experiment because the students were randomly assigned to the treatments – SAT review course and no SAT review course. Because the students were volunteers and were not randomly selected from the population of interest, the results of the experiment cannot be used to determine whether an SAT preparation course would be effective in improving the scores of all seniors in this school.

2. **(C)** The null hypothesis always involves an equality, and the alternative hypothesis always involves an inequality. Thus options (A) and (E) are incorrect, since the null hypothesis given in those two options is an inequality. Furthermore, the airline wishes to provide convincing evidence that the airplanes are, on average, *less than* 8 minutes late in landing. Thus, looking at the alternative hypotheses given, (B) and (D) are incorrect responses, and (C) is the correct response.

3. **(D)** Converting to z-scores using the z-score formula, $z = \frac{x - \mu}{\sigma}$, the z-scores would be:

   Test I: $\frac{600 - 500}{80} = 1.25$, Test II: $\frac{600 - 470}{120} = 1.08$, Test III: $\frac{600 - 560}{30} = 1.33$.

   The test with the highest z-score would be this student's best test. So, Test III was this student's best test, then Test I, and then Test II. The answer is: III, I, II.

4. **(D)** To assign the digits, you need to know the probability of getting any particular question correct. Since there are four possible answers, this probability is 25%. Choice (D) is the only choice that produces this correct probability. Choice (E) is not correct because 00 – 25 is 26 digits out of 100 or 26%.

5. **(E)** When rolling the pair of dice once, we can either get 0, 1, or 2 two's. We know that the probability of rolling a 2 on one of the dice is 2/6 = 1/3 and the probability of not rolling a 2 is 4/6 = 2/3. Using the multiplication rule for independent events, the probability distribution below shows the probabilities for 0, 1 or 2 two's in one roll of the two dice.

No. of 2's (x)	Probability, $p(x)$
0	$\left(\frac{2}{3}\right)\left(\frac{2}{3}\right) = \frac{4}{9}$
1	$2\left(\frac{1}{3}\right)\left(\frac{2}{3}\right) = \frac{4}{9}$
2	$\left(\frac{1}{3}\right)\left(\frac{1}{3}\right) = \frac{1}{9}$

$$E(x) = \sum x \cdot p(x) = 0\left(\frac{4}{9}\right) + 1\left(\frac{4}{9}\right) + 2\left(\frac{1}{9}\right) = \frac{6}{9} = \frac{2}{3}.$$

6. **(E)** In order for the formula for correlation, $r = \frac{1}{n-1}\sum\left(\frac{x_i - \bar{x}}{s_x}\right)\left(\frac{y_i - \bar{y}}{s_y}\right)$ to give us a value that is –1 times the original value, we need to *either* multiply all the x-values by a negative number *or* all the y-values by a negative number. Multiplying both the x's and the y's by –1 would give you a correlation of 0.3. Subtracting a value larger than $\bar{x}$ from each of the x-values is simply a linear transformation to the left, changing the y-intercept but not the correlation or the slope. Switching the x-values and y-values would not change the value of r: correlation makes no distinction between x's and y's. Subtracting the mean from both the x's and the y's produces a translation in both the x and y directions but does not change the correlation or the slope.

7. **(C)** The formula for the mean is: $\frac{\sum x_i}{n}$. If we multiply each of the x-values by k we get: $\frac{\sum k \cdot x_i}{n} = \frac{k\sum x_i}{n} = k\left(\frac{\sum x_i}{n}\right)$, and so the mean is multiplied by k. The formula for the variance is $s^2 = \frac{\sum (x_i - \bar{x})^2}{n-1}$. If we multiply each term by k we get:

$$s^2 = \frac{\sum (k \cdot x_i - k\bar{x})^2}{n-1} = \frac{\sum k^2 (x_i - \bar{x})^2}{n-1} = \frac{k^2 \sum (x_i - \bar{x})^2}{n-1}$$

, and so the variance is multiplied by k^2.

8. **(C)** The formula for a 95% confidence interval for the difference of two proportions is: $(\hat{p}_1 - \hat{p}_2) \pm z^* \sqrt{\frac{\hat{p}_1(1-\hat{p}_1)}{n_1} + \frac{\hat{p}_2(1-\hat{p}_2)}{n_2}}$. In this case $\hat{p}_1 = \frac{84}{210} = 0.4$ and $\hat{p}_2 = \frac{77}{220} = 0.35$. For the required confidence interval, this gives us:

$$0.05 \pm 1.96\sqrt{\frac{0.4 \times 0.6}{210} + \frac{0.35 \times 0.65}{220}}.$$

9. **(C)** The first step in solving this problem is to find the percentage the top 60 students constitute of the whole class. $\frac{60}{400} = 0.15$. Since 15% are the top 60, 85% have scores lower than that value. Use the table of normal probabilities to find the z-score that corresponds to an area to the left of 0.85. The closest value to 0.85 using the table is 0.8508. The z-score that corresponds to this area under the curve is $z = 1.04$. Use this value to solve for x in the equation: $z = \frac{x - \mu}{\sigma}$. Substituting the values of z and μ, we get: $1.04 = \frac{x - 80}{6}$. The solution to this equation is $x = 86.24$. Therefore the highest score exceeded by the top 60 students is 86. This problem can also be solved using the inverse normal feature of the graphing calculator. Press 2nd/VARS to get to the distribution menu. Choose 3:invNorm and execute the command. Fill in the parenthesis using, area to the left of x, mean, and standard deviation as follows: invNorm(.85,80,6). Then press ENTER.

10. **(A)** The formula is: Expected Count $= \dfrac{\text{(Row Total)(Column Total)}}{\text{Grand Total}}$. Because we are looking for the expected number of males who earned a grade of B, the row total is the total number of males (69), the column total is the total number of students earning a B (73), and the grand total is the total number of students (141). This gives us: $\dfrac{69 \cdot 73}{141}$ for the expected count.

11. **(C)** Students may confuse this example with difference of proportions ($H_0\!: p_m = p_f$, where p_f is the proportion of females and p_m is the proportion of males). This would be incorrect because the males and females are being selected from the same 60 people who are all SUV drivers. The proportion of males is simply 1 – the proportion of females. The difference of proportion test stated above requires that the two proportions come from different populations. Since we are asking whether SUV drivers are more likely to be male we must test the hypothesis that the proportion of males is equal to 50% (anything greater than 50% would be a majority) against the alternative hypothesis that the proportion of males is more than 50%. This requires a one-tailed test. Since the "large list" includes all SUV drivers and because a simple random sample was used to select students from this list, it would be appropriate to perform the test for a population proportion in this situation.

12. **(D)** The trade off for more confidence is a wider interval. The formula for a confidence interval for the mean is $\bar{x} \pm t^* \dfrac{s}{\sqrt{n}}$. For greater values of C (your confidence level) you get larger values of t^*. This creates wider intervals.

13. **(C)** The boxplot shows that the distribution of grades is right skewed and has a median of 70. Because the boxplot is right skewed, the mean will be pulled to the larger numbers supporting the statement that the mean will be greater than 70. In addition, because the median is 70, 50% of the students scored above this number and 50% of the students scored below this number. The interquartile range is represented by the width of the box, which is greater than ten.

14. **(B)** This is an experiment since the subjects are made to walk in particular ways (rather than being observed using their own walking styles); thus, statement I is true. It is a matched-pairs design since each participant is required to walk in both styles; for each student, the increases in pulse rate for the two types of walk form a pair. Thus statement II is true. Since it is a matched pairs design, the paired *t*-test should be used to analyze the results; a two-sample *t*-test would be incorrect. Thus, statement III is false, and the correct response is (B).

15. **(C)** The explanatory variable is the brand of paper towel. When attempting to determine what the explanatory variable is, ask yourself what it is whose effect on the response variable is being evaluated. In this case the student is comparing the "breaking time" between the brands. Thus "brand" is the explanatory variable.

16. **(B)** The residual = observed y – expected y. The observed final exam result for George is 85. The expected y-value for George is found by substituting 80, the observed first test score, for x in the least squares regression equation. This gives us $25 + 0.7(80) = 81$. The residual is found by subtracting this expected score from George's observed score. This gives us $85 - 81 = 4$.

17. **(D)** Statements I and III are true. The spread of the sampling distribution of the sample mean is given by $\frac{\sigma}{\sqrt{n}}$. Therefore, for larger values of n, the standard deviation of the sampling distribution will get smaller. Thus, statement I is true and statement II is false. Statement III is the Central Limit Theorem and therefore is true.

18. **(B)** The appropriate inference procedure to use in this problem is the χ^2 goodness of fit test.
    The appropriate set of hypotheses would be:
    H_0: $p_1 = 0.50, p_2 = 0.2, p_3 = 0.3$
    H_a: At least one of the proportions is different from the given value.
    The value of the test statistic can be found by using the formula:
    $$\chi^2 = \sum \frac{(\text{observed} - \text{expected})^2}{\text{expected}}$$
    - The observed number of students who attended college and obtained a degree was 510. The expected frequency for that category is $1000 \times 0.5 = 500$
    - The observed number of students who had attended college but not obtained a degree was 160 and the expected frequency for that category is $1000 \times 0.20 = 200$. This leaves $1000 - (510 + 160) = 330$ who had not attended college.
    - The value for χ^2 can be obtained by substituting the values of the observed and the expected frequencies into the formula:
    $$\chi^2 = \frac{(510 - 500)^2}{500} + \frac{(160 - 200)^2}{200} + \frac{(330 - 300)^2}{300} = 11.2$$
    - The p-value corresponding to this value of the chi-square statistic is 0.0037.

    Because the p-value is so small, we have reason to reject the null hypothesis and conclude that the distribution has changed.

    This problem can be done on the TI-84 calculator by using the list menu. Put the observed values of 510, 160, and 330 into L1. Put the expected values of 500, 200 and 300 into L2.

    Press STAT/ TESTS/D: χ^2GOF-Test, and press Enter. Press Calculate.

19. **(D)** First, find the area of the tails corresponding to the middle area under the normal curve equal to 0.97. This can be found by subtraction 0.97 from one and then dividing by two. $\frac{1-0.97}{2} = 0.015$.

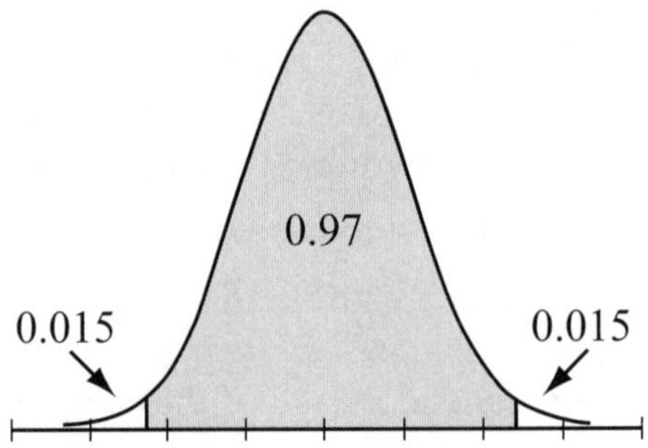

Using the negative standard normal probability table and looking up the area of 0.015, we find the corresponding value of z to be -2.17. (The positive standard normal probability table could also have been used. To use the positive table, look up $0.97 + .015 = 0.985$.) Next we use the formula: $n = p^*(1-p^*)\left(\frac{z^*}{m}\right)$ (where p^* is an estimated or guessed value for the sample proportion) to find the minimum sample size that ensures that the error will be no more than 3 percentage points. Because no information is about the true population percentage we will substitute $p^* = 0.5$. This gives us: $n = (.5)(.5)\left(\frac{-2.17}{0.03}\right)^2 = 1308.0278$. Rounding up, we get: $n = 1309$.

20. **(E)** The formula for the confidence interval for the slope is: $b \pm t^* s_b$. The appropriate degrees of freedom for calculating t^* is $n - 2 = 28$. Using the row of the t-distribution table for 28 degrees of freedom and an upper tail probability of $\frac{(1-.95)}{2} = .025$, we find the value of t^* to be 2.048. To complete the formula, use $b = 0.0000515$, and $s_b = 0.000005$.

    Don't confuse the value of s with that of s_b. The value of $s = 1.33$ is the standard error about the least squares line. This value is an estimation of σ which is a measure of variability in vertical direction between the expected value of y and the observed value of y for a given value of x. The value of s is also known as the estimated standard deviation of the residuals. The value of s_b is the value of the standard error of the sampling distribution of the slope.

21. **(E)** Choice (A) is incorrect because if the distribution were skewed left, the mean would be less than the median. Choice (B) is incorrect because the fact that the mean is greater than the median tells us that the distribution is positively skewed (that is, not symmetrical). Choice (C) is incorrect because 50 and 68 are the minimum and the first quartile respectively, respectively, and so approximately 25% (not 50%) of the data values fall between 50 and 68. Choice (D) is incorrect since data values that are within one standard deviation of the mean would lie between 62.91 and 97.09. Since 50% of the data values lie between the first quartile (68) and third quartile (87), then at least 50% (and therefore more than 40%) must lie between 62.91 and 97.09. Choice (E) is correct. Using the formula $Q_3 + 1.5IQR$ we get: $87 + 1.5(19) = 115.5$. Since 140 is larger than this number, there is at least one outlier.

22. **(E)** I and III are true. The null and alternative hypotheses for this test would be as follows:

    $H_0: p = 0.5$

    $H_a: p > 0.5$

    For small p-values we reject the null hypothesis. From our printout, the p-value is 0.0044. At the 0.01 level of significance, we reject the null hypothesis and conclude that there is sufficient evidence to conclude that the population proportion is larger than 0.50. Therefore statement I is true and statement II is false. Statement III is true because it is the definition of the p-value.

23. **(C)** A histogram for the twelve brands of cereal can be created from the cumulative frequency histogram. It is shown below. Seventy-five percent of the brands consist of 9 brands. Only five brands have at least 25 grams of carbohydrates.

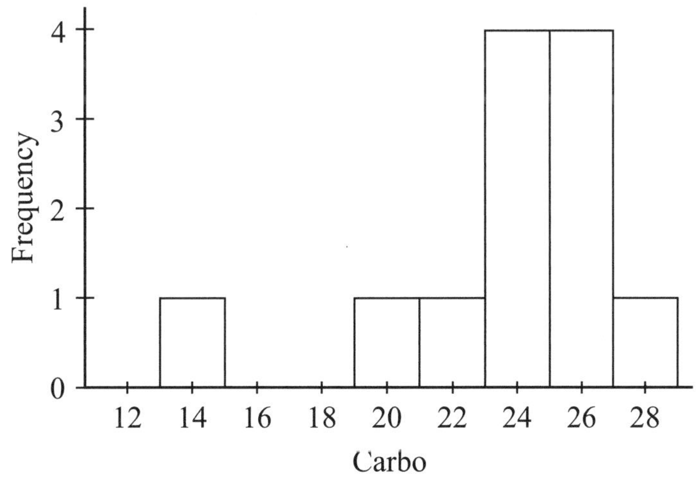

You can also use the original cumulative percent frequency histogram to determine that the statement in choice (C) is not correct (making choice (C) the correct answer). It appears from the graph that approximately 60% of the brands of cereal have less than 25 grams of carbohydrates, and so approximately 40% have more than 25 grams of carbohydrates. Choice (A) is a conclusion that can be reached from the graph. If the midpoint of each class is used to estimate the number of grams of carbohydrates then the six brands with the lowest amount of carbohydrates contain respectively: 14, 20, 22, 24, 24, and 24 grams. The six brands with the highest amounts of carbohydrates contain 24, 26, 26, 26, 26 and 28 grams of carbohydrates. The range of the six brands with the higher amounts of carbohydrates is 4 grams, and the range of the six brands with lower amounts of carbohydrates is 10 grams. Clearly the six brands with the lower amounts of carbohydrates have more variability than the six brands with higher amounts of carbohydrates. Choice (B) is also a correct conclusion. The median carbohydrate quantity is in the class containing cereals that have at least 23 and less than 25 grams of carbohydrates. Choice (D) represents a correct conclusion, also. Only about 25% of the cereals contain 23 or less grams of carbohydrates, and so most of the cereals would be off limits. Choice (E) is also a correct conclusion. When the bars of a cumulative percent histogram remain flat, this is indicating that no additional data has been added from the previous class.

24. **(D)** The power of the test increases as sample size increases, because increased sample size decreases the variability of the sample mean. This affects the power of the test because the power is defined to be 1 – *P* (Type II error). The smaller the probability of a Type II error, the larger the power of the test. Also, Type II errors depend on how far the true population mean is from the hypothesized mean. The further away the true mean is from the hypothetical mean, the smaller the probability of a Type II error and consequently the larger the power. Choice (D) has the largest sample size size and the value of μ that is furthest away from 10.

25. **(B)** Because there are only two treatments, old drug and new drug, we are blocking in groups of two. This is actually a matched pairs design which is a special case of blocking. The experimenter needs to control the influences of age and gender; therefore each block must be as closely matched as possible in terms of both age and gender. Each block needs to contain either two women or two men with the closest ages possible.

26. **(B)** The standard deviation of the sampling distribution for the difference of two means is: $\sigma_{\bar{x}_1 - \bar{x}_2} = \sqrt{\frac{\sigma_1^2}{n_1} + \frac{\sigma_2^2}{n_2}}$. That would give us: $\sqrt{\frac{3.3^2}{100} + \frac{2.6^2}{100}} = 0.42$

27. **(D)** To linearize an exponential function, you take the log of the *y*-values. If $y = ab^x$, then taking the log of the *y*-values will yield:

$\log y = \log(ab^x)$

$\log y = \log a + \log b^x$

$\log y = \log a + x \log b$

This creates a linear model where log *a* is the *y* intercept and log *b* is the slope of the line. The transformed *y* variable (log *y*) is the new response variable while *x* remains the explanatory variable.

28. **(C)** This is by the definition of confidence interval. A level C confidence interval for a parameter θ is an interval computed from sample data by a method that has probability C of producing an interval containing the true value of θ.

29. **(A)** Slope is defined to be change in the response variable divided by the change in the explanatory variable. In this case, the response variable is death rate and the explanatory variable is years. A slope of –3.63 must be interpreted as a predicted drop of 3.63 in death rate per one year.

30. **(B)** Statement I is false because reducing the probability of making a Type I error increases the probability of a Type II error. Statement II is true because the power of the test is 1 – probability of a Type II error for a given true alternative hypothesis. Statement III is false because the *p*-value is the probability of getting a test statistic as extreme as we did given that the null hypothesis is true.

31. **(D)** There are 34 heights that fall below 67 inches. Divide 34 by the total number of heights which is forty $\frac{34}{40} = 0.85$. We get that 67 is the 85th percentile.

32. **(D)** Points that lie on the least squares line that are not near $(\bar{x}, \bar{y})$ are influential and increase the value of r^2. The point (4,300, 15.63) lies on the line $y = 62.5 - 0.0109x$ and is not near $(\bar{x}, \bar{y})$.

33. **(A)** $P\,(\text{sum of seven/the first die is odd}) = \dfrac{P(\text{sum of seven and the first die is odd})}{P(\text{the first die is odd})}$

The probability we get a seven and the first die is odd can be computed using a sample space.

	1	2	3	4	5	6
1						1,6
2					2,5	
3				3,4		
4			4,3			
5		5,2				
6	6,1					

Only 3 out of 36 elements $\left(\frac{1}{12}\right)$ have a sum of seven and the first number is odd. The probability that the first die shows an odd number is $\frac{1}{2}$. This gives us a probability of:

$$\frac{\frac{1}{12}}{\frac{1}{2}} = \frac{1}{6}.$$

34. **(C)** In order to make a Type II error we will fail to reject the null hypothesis when it is false. The probability of the Type II error is the probability that the patient is diagnosed with the disease given that he does not have the disease. Under the condition that the patient does not have the disease, there were 15 times that the test indicated that the disease was present. This gives us a probability of 15/140.

35. **(E)** This is a sampling distribution for proportions problem. Asking the probability that fewer than 448 out of 1600 or $\frac{448}{1600} = 0.28$ or 28% are against the current administration policy, is the same as asking the probability that more than 72% are for the policies of the administration. The mean of the sampling distribution for $\hat{p}$ is p and the standard deviation is: $\sqrt{\frac{p(1-p)}{n}}$ where p is the true population proportion. The value of $p = 0.70$ if we want to find the proportion for the policies of the administration or $p = 0.30$ if we choose to find the proportion against the policies. Because the question specifically asks for the probability that less than 28% are against, but choices C, D and E are stated in terms of *more than*, our strategy will be to use $p = 0.7$ and find the probability that more than 72% are for the policies of the administration. The value of $\hat{p}$ is the proportion in our survey who are for the policies, which is 0.72. Using the formula: $z = \frac{\hat{p} - p}{\sqrt{\frac{p(1-p)}{n}}}$, we get: $P\left(z > \frac{0.72 - 0.7}{\sqrt{\frac{(0.7)(0.3)}{1600}}}\right)$ since we are looking for the probability that more than 72% are for the policies of the administration.

36. **(E)** A simple random sample is a selection method by which all possible samples of size n are equally likely to be selected. In choice A, not all samples of size 60 have the same chance of being selected since we are selecting all of the first 60 students to arrive at school. Students who do not arrive early have no chance of being selected. Choosing every tenth student (Choice B) to enter the library also does not allow every group of 60 students to have the same chance of being selected. Choice C mandates that the selected group must contain an equal number of first, second, third and fourth-year students. A sample consisting of 30 freshmen and 30 seniors has no chance of being selected and therefore is not compatible with the definition of a simple random sample. Choice D ignores all students that are not using the cafeteria. Choice E allows every group of 60 students to have the same chance of being selected.

37. **(A)** There is only one way to get a sum of a 2. We must get a 0 on die A *and* a 2 on die B. The probability of this is $\left(\frac{1}{6}\right)\left(\frac{2}{6}\right) = \frac{1}{18}$. This narrows our solution down to choice A or D. There are two ways of getting a sum of 3. We could get a 0 on die A *and* a 3 on die B *or* we can get a 1 on die A *and* a 2 on die B. The probability of this occurring is: $\left(\frac{1}{6}\right)\left(\frac{4}{6}\right) + \left(\frac{2}{6}\right)\left(\frac{2}{6}\right) = \frac{2}{9}$. This gives us choice A. Similar methods are used to check the rest of the probabilities. To obtain a sum of 4 we need either a 1 on die A *and* a 3 on die B *or* a 2 on die A *and* a 2 on die B. The probability of getting a 1 *and* a 3 is: $\left(\frac{2}{6}\right)\left(\frac{4}{6}\right) = \frac{2}{9}$. The probability of getting a 2 *and* a 2 is: $\left(\frac{3}{6}\right)\left(\frac{2}{6}\right) = \frac{1}{6}$. The probability of getting either one of these two events is: $\left(\frac{2}{9}\right)\left(\frac{1}{6}\right) = \frac{7}{18}$. To get a sum of 5 we need to get a 2 on die A and a 3 on die B. This probability is: $\left(\frac{3}{6}\right)\left(\frac{4}{6}\right) = \frac{1}{3}$. This agrees with choice A. Another solution strategy would be to create a sample space of the 36 equally likely events (sums of die A and die B) that could occur when tossing these two dice. The table below shows these events.

Die B

Die A		2	2	3	3	3	3
	2	4	4	5	5	5	5
	2	4	4	5	5	5	5
	2	4	4	5	5	5	5
	1	3	3	4	4	4	4
	1	3	3	4	4	4	4
	0	2	2	3	3	3	3

38. **(D)** Let X = the number of free cups of coffee that the customer obtains. X has a binomial distribution with $n = 5, p = \frac{1}{6}$ and $k = 0, 1, 2, 3, 4,$ and 5. The table below shows the theoretical probabilities and the expected counts if 100 trials were simulated.

k	$P(X = k)$	$100 \cdot P(X = k)$
0	.40188	40.2
1	.40188	40.2
2	.16075	16
3	.03215	3
4	.00322	0
5	.00013	0

To get the probabilities found in the table above, you can use the formula:

$P(X = k) = \binom{n}{k} p^k (1 - p)^{n-k}$ OR using the values for n, p and k given above, you can use your TI-84. Follow the steps below to use your calculator:

2nd DISTR⇒A:binompdf($5, \frac{1}{6}$). Press ENTER and scroll right to see all of the probabilities given in the table above. Simulations will frequently not follow the theoretical values exactly, but will come close. The closest to the values listed above would be answer (D).

Number of Cups of Coffee Free	Frequency
0	39
1	42
2	14
3	4
4	1
5	0

39. **(B)** $\frac{5}{6}\$1.00 + \frac{1}{6}\$0.00 - \$0.35 = \$0.48$. The probability a customer pays \$1.00 is $\frac{5}{6}$. The probability that a customer pays nothing is $\frac{1}{6}$. The expected value is $\sum x \cdot p(x)$. From this value we subtract the given cost of 35 cents to find the expected profit per cup of coffee.

40. **(B)** The area under the curve to the right of $z = 1.36$ is 0.0869. We can get this answer by looking up $z = 1.36$ in the tables, getting .9131, then, subtracting from 1 to get 0.0869. Double this, because it is a two tailed test, we get 0.1738. Using the TI-84, we can get the answer by using 2nd DISTR normalcdf(1.36, 1E16). This gives us 0.0869. Doubling this result because this is a two tailed test, we get 0.1738. Alternatively, realizing from the start this is two tailed, do 1 – normalcdf(–1.36,1.36) on the TI-84 calculator. This gives us: 0.1738.

Section II

Question One

(a) The median sodium content for sandwiches from restaurant "M" is just about 900 mg, while the median sodium content for sandwiches from restaurant "S" is about 1,250 mg. The interquartile range for restaurant "M" is smaller than 500 mg while the interquartile range for restaurant "S" is larger than 500 mg. The range for restaurant "M" is about 1,000 mg of sodium while the range for restaurant "S" is approximately 1,250 mg of sodium. The distribution of sodium content for restaurant "M" has less variability than the distribution of sodium content for restaurant "S". The boxplots of sodium content for both restaurants are roughly symmetrical. All of the sandwich options for restaurant "M" have less sodium content than twenty-five percent of the sandwich options for restaurant "S". Seventy-five percent (Q_3) of the sandwich options from restaurant "M" have less sodium content than fifty-percent of the sandwich options from restaurant "S". Fifty percent of the sandwich options for restaurant "M" have less sodium content than seventy-five percent of the sandwich options than restaurant "S". It appears that restaurant "M" has more choices with lower sodium content.

(b) The median mg. of sodium per gram for sandwich options at restaurant "M" is about 2.4, while for restaurant "S", the median is about 1.4 mg. of sodium per gram. Comparing the variability of the two restaurants, we find the interquartile range of sandwich options from restaurant "S" to be slightly over 0.5 while the interquartile range for sandwiches from restaurant, "M" is a little under 0.5 mg of sodium per gram. The range of mg of sodium for sandwich options from restaurant "M" is slightly over 1, while the range for sandwich options from restaurant "S" is slightly less than 1. There is not a great deal of difference in variability. Seventy-five percent of the sandwich options at restaurant "M" have more sodium per gram, than all of the sandwich options at restaurant "S" whereas seventy-five percent of all the sandwich options of restaurant "S" have less sodium per gram than all of the sandwich options as restaurant "M". The boxplot for restaurant "M" is roughly symmetrical. The boxplot for restaurant "S" is right skewed. Restaurant "S", generally speaking, has sandwich options with less sodium per gram than does restaurant "M".

(c) A statistical test to determine whether the mean sodium/g of all sandwich choices at restaurant "S" is different from the mean sodium/g of the all sandwich choices at restaurant "M" is inappropriate because neither of the data sets used comes from a random sample from the population of interest. As was stated in the stem of the problem, the sodium content for all of the sandwich options for each restaurants was recorded, which represents a population, not a sample. Inference procedures are inappropriate when we are using data from an entire population.

Question Two

(a) The regression equation is: Predicted Time = 15.027(Distance) – 5826. In the least squares line, *Time* denotes the time in days for a planet to revolve around the sun and *Distance* denotes distance of the planet from the sun in millions of kilometers.

Predictor	Coef	StDev	T	P
Constant	**-5826**	3303	-1.76	0.121
Distance	**15.027**	1.218	12.34	0.000

S = 7471 R-Sq = 95.6% R-Sq(adj) = 95.0%

The value of the slope and the y-intercept can be found in the computer printout. The y-intercept or constant is shown under the column "`Coef`" in row "`Constant`". The value of the slope is shown under the column "`Coef`" in row "`Distance`".

(b) The slope is interpreted as: for each increase of one million kilometer in distance that a planet is from the sun, it takes an estimated 15.027 additional days to revolve around the sun. Slope is defined to be change in y divided by change in x. Here y is the time in days for a planet to revolve around the sun and x is distance of the planet from the sun.

(c) Substituting 1,450 for Distance in the equation above, we get:
Predicted Time = 15.027(1450) – 5826 = 15963 days.

(d) The p-value of 0.000 in the Distance row indicates that there is strong evidence that a linear relationship exists between millions of kilometers the planet is from the sun and the number of days it takes to revolve around the sun. The p-value is the probability of getting a result at least as extreme as we actually obtained if the null hypothesis is true. That is, a p-value of .000 for the slope suggests that it is very unlikely to get the results that we did given that the null hypothesis is true. This leads us to reject the null hypothesis. The null hypothesis in this case would be $H_0: \beta = 0$, that is, the slope of the population regression line is 0 and there is no linear relationships between the explanatory and response variables.

Predictor	Coef	StDev	T	P
Constant	-5826	3303	-1.76	0.121
Distance	15.027	1.218	12.34	0.000

(e) The residual plot indicates that a line is not an appropriate model. There is a curved pattern to the residual plot. Our next step may be to try to make the data fit a straight line pattern by taking the logarithm of the y values, or the logarithm of both the x and the y values.

Question Three

(a) $\mu_{x-y} = 76.76 - 77.9 = -1.14$ seconds
$\sigma_{x-y} = \sqrt{\sigma_x^2 + \sigma_y^2} = \sqrt{(3.08)^2 + (2.49)^2} = 3.96$ seconds

(b) For the second swimmer to win the race, the time for the second swimmer (Y) must be less than the time for the time for the first swimmer (X). In other words, the first swimmer's time – second swimmer's time > 0, i.e. $X - Y > 0$. $X - Y$ follows a normal distribution with $\mu_{X-Y} = -1.14$ seconds and $\sigma_{X-Y} = 3.96$ seconds. So $P(X - Y > 0) = P\left(Z > \frac{0-(-1.14)}{3.96}\right) = P(Z > .29) = 0.3858$. Rounded to two decimals places, there is a 39% chance that swimmer two beats swimmer one in the race. The result could have been obtained by using normalcdf on the TI-84 graphing calculator. Press: 2nd distribution, and fill in the parenthesis as shown: normalcdf(0,1E16, -1.14, 3.96).

(c) The swimmer's times in any race are independent of their times in any other race. The probability that the second swimmer will beat the first swimmer in any race is 0.39. There is a fixed number of trials, $n = 10$. We want the probability of 1 or more successes. It will be easier to find the probability that second swimmer will never beat the first swimmer and subtract that answer from one. Let $X =$ the number of successes in ten races. Let $p = 0.39$; $n = 10$; $x = 0$ and use the binomial distribution probability formula: $P(X = k) = \binom{n}{k} p^k (1-p)^{n-k} = \binom{10}{0} 0.39^0 (1 - 0.39)^{10} = 0.007$ or 0.7% chance that the second swimmer will never beat the first swimmer. $1 - 0.007 = 0.993$ or 99.3% chance that the second swimmer will beat the first swimmer in at least one race. Using the TI-84 calculator, press 1-2nd Vars B:binomialcdf(*n*, *p*, *k*). Fill in the parenthesis as follows: binomialcdf(10, 0.39, 0) = 0.993.

Question Four

(a) On the basis of an assumption that the weights of all 18-year olds are normally distributed, the expected frequencies can be calculated as follows:

Given that $\mu = 148$ pounds and $\sigma = 15$,

$P(118 < X < 148) = P\left(\frac{118 - 148}{15} < Z < \frac{133 - 148}{15}\right) = P(-2 < Z < -1)$. To find the probability that Z is between -1 and -2, we subtract the areas under the curve.

$P(Z < -1) - P(Z < -2) = 0.1587 - 0.0228 = 0.1359$.

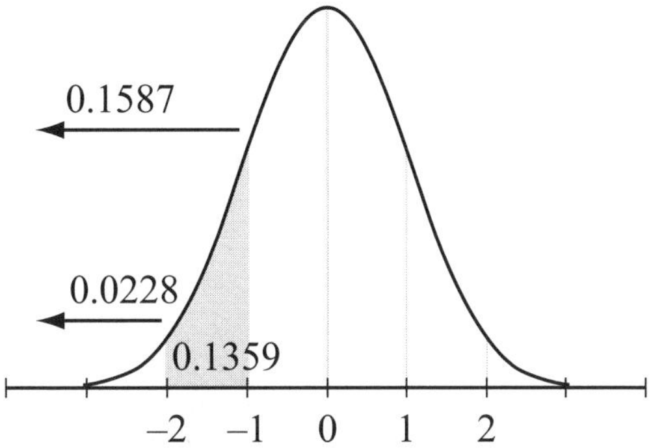

The expected frequency in a random sample of 300 18-year olds who weigh between 118 pounds and 133 pounds would be $(0.1359)(300) = 40.77$.

Alternatively, this answer could have been obtained by using the fifth class of the frequency distribution and applying the symmetrical properties of the normal curve.

	Weight (x)	Expected Frequencies
	$x \leq 118$	6.825
	$118 \leq x < 133$	
	$133 \leq x < 148$	
	$148 \leq x < 163$	
5th class →	$163 \leq x < 178$	40.77
	$x \geq 178$	6.825

The weight of 163 pounds is one standard deviation above the mean: $148 + 1(15) = 163$. The weight of 178 pounds is two standard deviations above the mean: $148 + 2(15) = 178$. The expected frequency for weights of 18-year-olds between 163 lbs and 178 lbs is given as 40.77. The expected frequency of 18-year-olds between the weights of 118 lbs. and 133 lbs is also 40.77 because these values are the same number of standard deviations below the mean as 178 and 163 are above the mean.

To compute the expected frequency of the third class, we find $P(133 < X < 148) = P\left(\frac{133 - 148}{15} < Z < \frac{148 - 148}{15}\right) = P(-1 < Z < 0)$. The expected frequency in a random sample of 300 18-year olds who weigh between 133 lbs. and 148 lbs. is $300(0.3413) = 102.39$. Calculations for the fourth class are exactly the same because 163 is one standard deviation above the mean. The expected counts are listed in the table below:

Weight (x)	Expected Frequencies
$x \leq 118$	6.825
$118 \leq x < 133$	40.77
$133 \leq x < 148$	102.39
$148 \leq x < 163$	102.39
$163 \leq x < 178$	40.77
$x \geq 178$	6.825

Or using the TI-84

To calculate the expected frequency in a random sample of 300 eighteen-year-olds who weigh between 118 lbs. and 133 lbs. Press: 2nd VARS 2:normalcdf(118, 133, 148, 15). To calculate the expected frequency of eighteen-year-olds who weigh between 133 lbs. and 148 lbs. Press: 2nd VARS 2:normalcdf(148, 163, 148, 15)

Or using the Empirical Rule

The Empirical Rule states that 68% of the values lie within one standard deviation of mean, 95% of the values lie within two standard deviations of the mean and 99.7% of the values lie within three standard deviations of the mean.

Note, again, that 163 and 133 are one standard deviation above and below the mean respectively. To calculate the expected frequency for the third or the fourth classes, divide 68% by 2. That tells us that 34% lie within one standard deviation of the mean. Multiplying this by 300, we get an expected frequency of 102.

To use the Empirical Rule to calculate the expected frequencies of the second class, note, again, that 133 is one standard deviation below the mean of 148 and 118 is two standard deviations below the mean. Divide both 95% and 68% by two to get 0.475 and 0.34. Subtracting these values, we find that 13.5% of the values lie between one and two standard deviations below the mean. Multiplying this by 300 we get 40.5.

(b) To determine whether the observed data shown give significant evidence to conclude that the weights of the population of 18-year-olds are not normally distributed with a mean of 148 lb and a standard deviation of 15 lb, we will conduct a chi-square goodness-of-fit test.

Part 1: Hypotheses:

H_0: $p_1 = 0.0227; p_2 = 0.1359; p_3 = 0.3413; p_4 = 0.3413; p_5 = 0.1359; p_6 = 0.0227$

H_a: At least one of these proportions is incorrect.

Part 2: Name of test and check of conditions:

Chi-square test of goodness-of-fit. $\chi^2 = \sum \frac{(\text{observed} - \text{expected})}{\text{expected}}$

- The sample is a simple random sample from the population of 18-year olds. Stated in the stem of the problem.
- All expected cell counts are at least 1 and no more than 20% of expected cell counts are less than 5. Since all expected cell counts are greater than 5, we can proceed with the test.

Weight (x)	Expected Frequencies
$x \leq 118$	6.825
$118 \leq x < 133$	40.77
$133 \leq x < 148$	102.39
$148 \leq x < 163$	102.39
$163 \leq x < 178$	40.77
$x \geq 178$	6.825

Part 3: Mechanics:

$$\chi^2 = \frac{(7 - 6.825)^2}{6.825} + \frac{(38 - 40.77)^2}{40.77} + \frac{(100 - 102.39)^2}{102.39} + \frac{(102 - 102.39)^2}{102.39}$$

$$+ \frac{(40 - 40.77)^2}{40.77} + \frac{(13 - 6.825)^2}{6.825} = 5.85.$$

Calculation of the p-value: Using the Chi-square table and $6 - 1 = 5$ degrees of freedom (number of categories – 1), the smallest value of χ^2 shown on the table is 6.63. The p-value associated with $\chi^2 = 6.63$ is 0.25. The value we calculated for χ^2 was 5.85. Since this value is less than 6.63 it means that the p-value associated with this value of chi-square is larger than 0.25.

Or using the calculator to obtain the p-value, $P(\chi^2 > 5.85)$, use 2nd distribution, χ^2cdf(5.85,1E19,5). This gives us a p-value of 0.32.

Part 4: Conclusion:

Because our p-value is greater than .05, we do not have sufficient evidence to reject the null hypothesis. We do not have convincing evidence that the weights of all eighteen-year-olds are not normally distributed with mean 148 lb and standard deviation 15 lb.

Question Five

(a) To answer this question, we need to perform a hypothesis test for the difference of two means based on independent samples.

Part 1: Hypotheses:

$H_0: \mu_{new} = \mu_{old}$ The mean improvement in systolic blood pressure for people who use the two drugs is the same.

$H_a: \mu_{new} > \mu_{old}$ The mean improvement in systolic blood pressure for people who use the new drug is greater than the mean improvement in systolic blood pressure for those who use the old drug.

Part 2: Identification of test by name or formula and checking appropriate conditions:
Name of Test: Two-Sample t test.

Formula: $t = \dfrac{\bar{x}_1 - \bar{x}_2}{\sqrt{\dfrac{s_1^2}{n_1} + \dfrac{s_2^2}{n_2}}}$ with degrees of freedom equal to the smaller of $n_1 - 1$ and $n_2 - 1$.

Conditions and test of conditions:

- Simple Random Sample. Have participants been randomly selected from the population of interest? It is stated in the stem of problem that they were. Random assignment of participants to treatments, old and new drug, will help ensure that differences in improvement in systolic readings between the two groups at the end of the study are due to the differences in drugs.
- Normality. Looking at the boxplots given in the problem we can safely assume that the distributions of improvement for both the old drug and the new drug were approximately normal since the boxplots are roughly symmetric and there are no outliers. (For the boxplot representing the new drug, the IQR is approximately $45 - 33 = 12$. 1.5 IQR would be calculated at 18 and that would put the fences at $45 + 18 = 63$ and $33 - 18 = 15$. No data points are above 63 or below 15. For the old drug, the IQR is again about 12. The fences are located at $40 + 18$ and $28 - 18$ or 58 and 10. Again no data points lie above the upper fence or below the lower fence.) Also, the sum of our two samples sizes is 24. For sample size greater than 15, two sample t procedures can be used except if strong skewness or outliers are present. Neither one of these conditions are present. Note also that symmetry could be argued for both data sets, as the mean and median are very close. For the new drug data set the median is between 35 and 40 and the mean is 38.75. For the old drug the mean is 33.25 and the median between 30 and 35. Also, since the two sample sizes are the same, t procedures are more robust against non-normality. Although normality cannot be established, it is appropriate to use t procedures in this case.
- Independent random samples / Random assignment. Since the subjects were randomly assigned to the two groups, we are justified in proceeding with the two-sample t-test.

Part 3: Mechanics:

Calculations: Let $\bar{x}_1 = 38.75$ (mean improvement of the new drug) and $\bar{x}_2 = 33.25$ (mean improvement of the old drug). What was given in the computer printout was the "Std Error Mean" which is the standard error of the mean. This value is defined to be $\frac{s}{\sqrt{n}}$. If we square the standard error values we get $\frac{s^2}{n}$. The square of this value is what we need in the formula for t which is; $t = \frac{\bar{x}_1 - \bar{x}_2}{\sqrt{\frac{s_1^2}{n_1} + \frac{s_2^2}{n_2}}}$. We will substitute 2.7 for $\frac{s_1}{\sqrt{n_1}}$ and 2.5 for $\frac{s_2}{\sqrt{n_2}}$. The formula for the test statistic, t, becomes: $t = \frac{38.75 - 33.25}{\sqrt{(2.7)^2 + (2.5)^2}} = 1.49$.

The p-value corresponding to the test statistic of $t = 1.49$ and degrees of freedom of $12 - 1 = 11$ is: $0.05 < p < 0.10$.

An exact p-value could also have been obtained using the TI-84 graphing calculator. Access the distribution menu, scroll down to 6: tcdf(1.49, 1E19, 11) = .082.

Part 4: Conclusion:

At the 0.5 level of significance, we cannot reject the null hypothesis. As the p-value of 0.082 is not less than the 0.05, there is insufficient evidence to suggest that there is a difference between the mean improvement in systolic blood pressure for people using the old drug and the new drug.

(b) Using the two scatterplots given, it is clear that there is a linear relationship between improvement in systolic blood pressure and the exercise level of the participant. As exercise level increases, improvement in systolic pressure generally increases. Level of exercise seems to affect the drugs' ability to improve systolic blood pressure. Blocking by exercise level would reduce variability. Two possible blocking arrangements are matched pair or a randomized complete block design.

To implement the matched pair design, block in groups of two by exercise level, matching exercise level for each pair as closely as possible. Continue this pairing until all participants are paired. Using a coin or a random integer generator, randomly select one member of each pair to get the old drug; the other member of the pair gets the new drug. The difference in improvement of systolic blood pressure for each pair of subjects is calculated at the end of six weeks.

A randomized block design would group the participants into three groups according to exercise level 0, 1 or 2. Within each block, subjects are randomly assigned to the two treatments, new drug or old drug. This could be done by assigning each participant with exercise level of 0 a number and using a random integer generator to randomly assign half of the group to get the old drug and the remaining half to get the new drug. Continue this process for each of the other two groups (exercise level 1 and exercise level 2). At the end of six weeks, the improvement in systolic blood pressure would be compared between the two treatment groups.

Section II
Part B

Question Six

(a) The probability of a randomly selected donor having type B or O blood is P(B) + P(O) = 0.15 + 0.45 = 0.60. Using the probability of 0.60 we can assign six of the ten integers to be a B or O blood type and assign four to be some other blood type. One possible solution is to assign 0 through 5 to represent blood type B or O and the remaining digits to represent not having blood type B or O. We then need to use the random number table and count how many digits it takes to arrive at three values between 0 and 5. This would tell us how many donors are needed to get three donors with blood types B or O.

(b) Starting in the upper left hand corner we use 8, **4**, **1**, 7, 7 and **0**. We stop at zero because we have found three values (4, 1, and 0) that are between 0 and 5 and representing three donors with type B or O blood. It took 6 donors to get three with the desired blood types. Beginning again where we left off, we use the numbers 6, 7, **5**, 7, **1**, 7, 6, and **1**. We stop at one because we have found three integers between 0 and 5. This time it took eight donors before finding three with the desired blood types. For the last trial we again start where we left off. The numbers begin with **3**, **1** and **5**. All three of these numbers are integers between 0 and 5; therefore we have three donors with type B or O blood in our first three numbers. Three trials of our simulation found that it took six, eight and three donors respectively, to get three donors with blood types B or O.

(c) Distribution Y is the correct distribution. Because the probability of getting a donor with type B or O blood is greater than the probability of finding a donor with type A or AB blood, it would take fewer donors to find three donors with type B or O blood type. Distribution Y has a lower mean and a lower maximum as well.

(d) Using distribution Y, the expected number is found using the formula: $E(X) = \sum x \cdot p(x)$. The values of $p(x)$ can be found by taking the frequencies for each x-value of the histogram and dividing by the sum of the frequencies, which is 100. The values of x are the values on the horizontal axis of the histogram. This gives us: $E(X) = 0.22(3) + 0.27(4) + 0.21(5) + 0.15(6) + 0.07(7) + 0.05(8) + 0.02(9)$ $+ 0.01(11) = 4.87$

(e) Blood bank C would be more likely to have a percentage of type O or B donors that exceeds seventy-five percent. This stands to reason since there are fewer donors at Blood Bank C and the Law of Large Numbers implies that the larger sample (Blood Bank D) is more likely than the other sample to have a sample proportion that is close to the true population proportion of 0.60. The higher variability in Blood Bank C's distribution is caused by the smaller sample size. With the higher variability, the chances of obtaining a sample percentage of 75% or more are higher for Blood Bank C.

This can also be explained using computational methods. If we treat the donors each day as a sample, the distribution of the proportion of type O and B donors for blood bank C would be approximately normal with a mean of 0.60 and a standard deviation of $\sqrt{\frac{0.6(0.4)}{50}} = 0.069$. The distribution of the proportion of type O and B donors for blood bank D has a mean of 0.60 and a standard deviation of $\sqrt{\frac{0.6(0.4)}{200}} = 0.035$.

We can find the appropriate probabilities by using the standard normal table since $n(p)$ and $n(1-p)$ are both at least 10. The z-score for blood bank C would be $\frac{0.75 - 0.60}{0.069} = 2.17$. The z-score for blood bank D is $\frac{0.75 - 0.60}{0.036} = 4.29$. The area under the normal curve to the right of 2.17 is 0.015. The area under the curve to the right of 4.33 is $8.9(10)^{-6}$.

Or, using the TI-84 calculator to calculate the probability for blood bank C, press: 2nd distribution and select normalcdf(0.75, 1E19, 0.6, 0.069) = 0.0149. To use the TI-84 calculator to calculate the probability for blood bank D, press: 2nd distribution and select normalcdf (0.75, 1E19, 0.6, 0.035) = .00000911. The probability of having at least 75% type B or O donors is greater for blood bank C than for blood bank D.

Sample Examination Three – Answers

1. **(D)** Adding and subtracting the margin of error from the sample proportion gives us the actual 95% confidence interval of (0.72, 0.84). A 95% confidence interval insures us that 95% of the intervals calculated from the same population and using the same sample size, will contain the population proportion. There is a 5% chance that the interval will not contain the population proportion. It is therefore reasonable to conclude that it is unlikely that the given sample result would have been obtained unless the true population proportion was between 0.72 and 0.84.

2. **(A)** The consumer group wishes to obtain convincing evidence that the program's claim is false, in other words that the mean weight loss per week is less than 4 pounds. Therefore the alternative hypothesis should be that $\mu < 4$. Consequently, the null hypothesis should either be that $\mu = 4$ or that $\mu \geq 4$. (Either of these would be acceptable, though the first form is the one more commonly used.) The second of these is not available in the options given, and so the correct response is (A).

3. **(C)** $3,000(0.4) = 1,200$ are full-time students and consequently 1,800 are part-time students. Seventy-five percent of the full-time students are education majors: $0.75(1,200) = 900$. Forty percent of the part-time students major in education: $0.4(1,800) = 720$. So, $(720 + 900)/3000$ or 54% major in education. If 200 students are randomly selected, we would expect $0.54(200) = 108$ to major in education.

4. **(B)** Using the formula: $b_1 = r\frac{s_y}{s_x}$ and other information from the printout, we get: $r = \sqrt{0.689} = 0.83$; $s_y = 2.302$ and $s_x = 3.610$. The response variable y is height because height depends on age. By our formula, we get: $b_1 = 0.83\left(\frac{2.302}{3.61}\right) = 0.53$.

5. **(C)** r^2 is the percent of the variation in y that is explained by the least-squares regression of y on x. Looking at the computer printout, we find that $r^2 = 68.9\%$. This is approximately 69%.

6. **(B)** The probability that a computer lasts less than three years is $0.12 + 0.02 + 0.05 = 0.19$. Therefore, the number of computers X out of the three that last less than three years has a binomial distribution with $n = 3$ and $p = 0.19$. Hence $P(X = 2) = \binom{3}{2}(0.19)^2(0.81)^1 = 3(0.19)^2(0.81)$, and so (B) is the correct response.

7. **(B)** The standard deviation is smaller as you get more values closer to the mean.

8. **(E)** The distributions are all symmetric and therefore all will have the mean equal to the median.

9. **(E)** First we need to know that inference for the slope requires $n - 2$ degrees of freedom. Since $n = 30, df = 30 - 2 = 28$. Inference for the slope uses the t-distribution. Using the 28 degree of freedom row in the t-table, we find two numbers that enclose our test statistic of 2.046. Since 2.046 is between 1.701 and 2.048, our p-value for a one-sided test would be between 0.025 and 0.05. Because our test is a two-tailed test, we need to double these numbers, giving us that the p-value lies between 0.05 and 0.10.

Tail Probability	.25	.20	.15	.10	.05	.025	.02	0.1	.005	.0025	.001	.0005
df = 28	0.683	0.855	1.056	1.313	1.701	2.048	2.154	2.467	2.763	3.047	3.408	3.674

10. **(D)** The batch will be accepted if at least 2 out of the 3 items are perfect. We will use the binomial formula which is $P(X = k) = \binom{n}{k} p^k (1 - p)^{n-k}$, where $p = 0.8, n = 3$ and $k = 2, 3$. $P(X = 2) = \binom{3}{2}(0.8)^2(0.2)^1 = 0.384$ and $P(X = 3) = \binom{3}{3}(0.8)^3(0.2)^0 = 0.512$. The probability that at least 2 are perfect is $0.384 + 0.512 = 0.896$.

Or, using TI-84 calculator, use: 2nd distribution; B: binomcdf(3, 0.8, 1). Subtract that answer from 1.This is the same as subtracting the sum of $P(X = 0)$ and $P(X = 1)$ from 1.

11. **(B)** Statement II is the definition of a simple random sample. Statement I is a property of a simple random sample but cannot stand alone as the definition. To better understand look at the following example. Suppose you have a room containing 10 boys and 10 girls. You wish to choose 2 students by randomly selecting one boy and one girl. In this example, each person has the same chance of being selected which is 0.1, but all groups of two students do not have the same probability of being selected. The probability of selecting two girls is zero. This is an example of a stratified random sample. Statement III is also not true for the same reason. This method of sampling does not allow all groups of 4 students to have the same chance of being selected.

12. **(B)** The significance level of a hypothesis test is the probability that a Type I error occurs.

13. **(C)** The method of instruction seems to depend on the type of class. Math and science classes do not use lecture to the extent that the art classes do. Also, using a test with a null hypothesis of "no association", we find the *p*-value is 0.017, which provides convincing evidence of an association between type of class and type of instruction.

    Or, using the TI-84 calculator, enter the data into matrix A. Press STAT, scroll down to TESTS and choose C: χ^2-TEST. Press ENTER. Scroll down to Calculate and press ENTER.

14. **(A)** Ninety-five percent of the data lie within two standard deviations of the mean. Using the formula: $\bar{x} \pm 2s$, we get: $22.3 + 2(5.4) = 33.1$ and $22.3 - 2(5.4) = 11.5$.

15. **(E)** For any given student, let the event that the student takes AP Statistics be *S* and the event that the student plays football be *F*. Since the two events are independent, the probability that a randomly chosen student does *both* of the two things is $P(S \cap F) = (1/7)(1/4) = 1/28$. The general addition rule for two events states that, for any two events *A* and *B*, $P(A \cup B) = P(A) + P(B) - P(A \cap B)$. So here, when a student is chosen at random,

$$\begin{aligned} P(\text{the student is involved in at least one of the two pursuits}) &= P(S \cup F) \\ &= P(S) + P(F) - P(S \cap F) \\ &= \frac{1}{7} + \frac{1}{4} - \frac{1}{7} \cdot \frac{1}{4} \\ &= \frac{4}{28} + \frac{7}{28} - \frac{1}{28} \\ &= \frac{5}{14}. \end{aligned}$$

16. **(B)** The first die has a one painted on one face and five blank faces. The probability of getting the 1 is $\frac{1}{6}$ and the probability of getting a blank face is $\frac{5}{6}$. We will treat the blank face as if it were a 0 since it adds no value to the sum. The expected value of this die is, $\sum x \cdot f(x) = \frac{1}{6}(1) + \frac{5}{6}(0) = \frac{1}{6}$. The second die has a two painted on one face and five blank faces. The expected value of the second die is, $\frac{1}{6}(2) + \frac{5}{6}(0) = \frac{2}{6}$. The third die has a three painted on one face and five blank faces. The expected value of the third die is, $\frac{1}{6}(3) + \frac{5}{6}(0) = \frac{3}{6}$. The expected values of the remaining three dice can be calculated in the same manner. The expected value of the sum of all six dice is: $\mu_{A+B+C+D+E+F} = \mu_A + \mu_B + \mu_C + \mu_D + \mu_E + \mu_F$. Substituting the expected values of each of the six dice, we get: $\mu = \frac{1}{6} + \frac{2}{6} + \frac{3}{6} + \frac{4}{6} + \frac{5}{6} + \frac{6}{6} = 3.5$.

17. **(D)** Since the residuals tend to be of a larger magnitude for low midterm scores than for high midterm scores, we can conclude that the points in the original scatterplot were generally further from the regression line for students with low midterm scores than for students with high midterm scores.

18. **(C)** $Z = \frac{x - \mu}{\sigma} \rightarrow 0.84 = \frac{x - 68}{2.5}$. Solving for x, we get: $x = 70.1$ (We use $z = 0.84$ because it is the z-score with a probability value closest to 80%.)

    OR using the TI-84 we go to: 2nd DISTR $\Rightarrow$ 3: Invnorm(0.8, 68, 2.5) = 70.1.

    We use 0.8 in place of 0.2 because we need to use the area from the left extreme of the curve to find the value of x. (see diagram below)

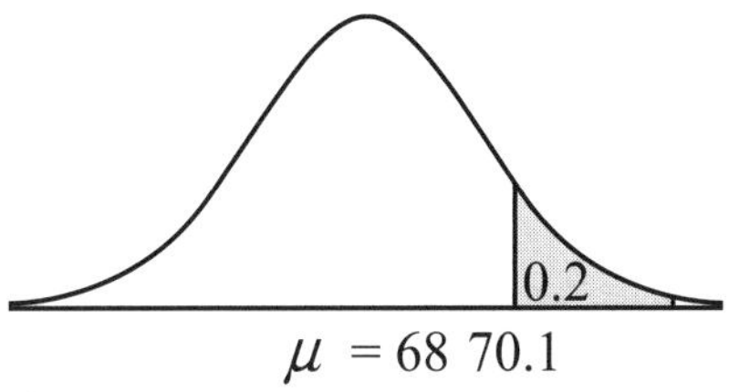

19. **(E)** Ages of students are likely to follow a uniform distribution. Students in high school vary from 15 to 18 with each age group likely to have about the same frequency.

20. **(B)** Degrees of freedom is (number of rows $-1) \times$ (number of columns $-1) =$ $(2 - 1)(3 - 1) = 2$.

21. **(C)** 150 lies further than 3 standard deviations above the mean. The value that is exactly three standard deviations above the mean is 130.

22. **(D)** Since the results of the five rolls of a die are independent, we can add the means and the variances of the results of the individual rolls in order to find the mean and the variance of the total score. To find the mean of the sum of the five rolls we can either add 3.5 five times or multiply 3.5 by five, giving us 17.5. The standard deviation of the score on a single roll is 1.5, which tells us that the variance of the score on a single roll is $1.5^2 = 2.25$. Multiply this by 5 to get 11.25 and then take the square root to convert variance back to standard deviation. This gives us 3.35. The formula used was:

$$\mu_{X_1+X_2+X_3+X_4+X_5} = \mu_{X_1} + \mu_{X_2} + \mu_{X_2} + \mu_{X_4} + \mu_{X_5} = 5\mu_X = 5(3.5) = 17.5.$$

    The standard deviation was found by applying this formula:

$$\sigma_{X_1+X_2+X_3+X_4+X_5} = \sqrt{\sigma^2_{X_1} + \sigma^2_{X_2} + \sigma^2_{X_3} + \sigma^2_{X_4} + \sigma^2_{X_5}} = \sqrt{5\sigma^2_X} = \sqrt{5(1.5)^2} = 3.35$$

23. **(C)** The 10th percentile has a z-score of -1.28. Shown below is the piece of the standard normal table applicable to this problem. 0.1003 is highlighted because it is the number closest to 10% or 0.10. Looking to the left and above, the z-score associated with that percentile is -1.28.

z	.00	.01	.02	.03	.04	.05	.06	.07	.08	.09
−1.2	.1151	.1131	.1112	.1093	.1075	.1056	.1038	.1020	.1003	.0985

That means that the value of the 10th percentile is 1.28 standard deviations below the mean. Since the standard deviation is given as 2.25, we multiply 2.25 by 1.28 to get 2.88 inches below the mean.

24. **(B)** The sum of the residuals is always 0. If $r^2 = 0$ then $r = 0$. Using the formula for the slope, $b_1 = r\frac{s_y}{s_x} = 0 \cdot \left(\frac{s_y}{s_x}\right) = 0$, so the line is horizontal.

25. **(D)** The probability of success, p, is the probability of getting either a 0 or a 1 from the random digits 0, 1, 2, 3, and 4. Since that is 2 out of 5, $p = .40$. Choices (A), (D) and (E) are still possible. There are 200 runs of the simulation. In each run, 30 random numbers are generated. The results of these 30 trials are independent, and each trial is a success or a failure (either we get a 0 or a 1 or we do not), there are a fixed number of trials ($n = 30$), and for each trial, the probability of success remains the same, ($p = 0.4$). From this we see that the distribution that is being simulated is a binomial distribution, with $n = 30$ and $p = 0.4$.

26. **(B)** The computer printout shown in the problem and below gives us two pieces of information. It gives a 95% confidence interval and it gives us a p-value for a one-tailed test. The 95% confidence interval shown below is not helpful in a one-tailed test. The p-value for a one-tailed test is what we need to use. Since that p-value is less than .05, the results of the test are significant at the .05 level. We can reject the null hypothesis that the two population means are the same and conclude that the mean at rest pulse rate for women is less than the mean at rest pulse rate for men at the .05 level of significance.

```
95% CI for μ(F)- μ(M):(-9.3,0.6)      ← confidence interval
T-Test μ(F) = μ(M)(vs <):
T = -1.79 P = 0.042 DF = 28           ← p-value using a one-tailed test
```

27. **(E)** The sampling distribution of the mean has mean equal to the mean of the population and standard deviation equal to $\frac{\sigma}{\sqrt{n}}$. The Central Limit Theorem states that for large sample sizes, the sampling distribution for the mean is approximately normal regardless of the shape of the population distribution. We need to pick the distribution that is approximately normal, with a mean of 12 and standard deviation of $\frac{\sigma}{\sqrt{n}}$ or $\frac{9.6}{\sqrt{100}} = 0.96$. We know that in a normal distribution, 99.7% of the values lie within 3 standard deviations of the mean. That gives us $(3)(0.96) = 2.88$. Add and subtract this value from the mean of 12.3, and we get 9.42 to 15.18. The problem states that there were 1000 trials. That means about $0.997(1000)$ or 997 of the values should be in that interval. The histogram that comes the closest to having about 997 values in the stated interval is the histogram in (E).

28. **(C)** If the p-value of the test statistic is 0.074, because it is a two tailed test it means that the area under the curve shown below (the distribution of the difference in the sample means) to the right of $(\bar{x}_1 - \bar{x}_2)$ is $0.074/2 = 0.037$. That means that the area to the left of $(\bar{x}_1 - \bar{x}_2)$ and to the right of 0 (the mean of the distribution) is $0.5 - 0.037 = 0.463$. See the diagram below. The smallest level of confidence for which the interval would contain 0 would have to be double 0.463, that is, 92.6%. The first level of confidence larger than this in our options is (C), 93%.

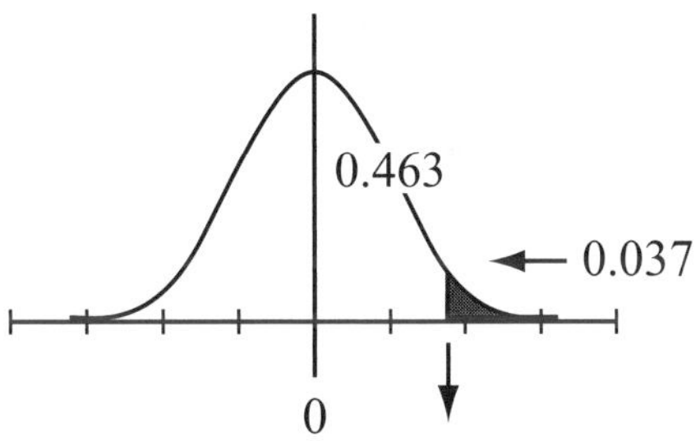

29. **(A)** Since there are 41 pieces of data, the median is in the 21st position. This means it is located in the interval representing the years in age of at least 10 but less than 15 years. Since in choice (E) the median is larger than 15, we can eliminate choice (E). The maximum data point has to be between 35 and 39 years old. This information is important as it eliminates choice (C) which shows a maximum of less than 35 years. The third quartile of the data is found by finding the middle value of the second half of the data. There are 20 numbers in the second half of the data so the third quartile is half way between the 31st and the 32nd data values. That would be in the interval of at least 15 years but less than 20 years old. This information eliminates choice (B). Choice (B) indicates that the third quartile is slightly above 20. The first quartile is between the 10th and the 11th position. The first quartile is located in the interval of at least 0 but less

than 5 years of age. This information eliminates choice (D). Choice (D) shows a first quartile that is between 5 and 10 years old. This leaves choice (A).

30. **(C)** Since the organizers select random samples from each of the sets of instrumentalists, and put the samples together to form the required sample, this is stratified random sampling. This should not be confused with *cluster* sampling. If, for example, each high school band involved in the gala was considered to be representative of the whole gathering in terms of opinion, then the organizers could select a random sample of high school bands, and interview the students in those bands. This would be cluster sampling. The sampling would be *systematic* if, for example, every tenth musician in the list of all musicians involved in the gala were selected.

31. **(C)** A paired t-test is the appropriate test in this case, since the data values come in pairs, one pair for each married couple. Neither a two sample z-test nor a two sample t-test would be appropriate as the data were not obtained by means of independent samples. A chi-square test of homogeneity is not correct because this is quantitative data not categorical data. The last option, a linear regression t-test has no relevance to the question that is being asked.

32. **(D)** There is no need for a control group in this design, since you are merely comparing two treatments. Choice (A) represents a flaw in the design of the experiment because students attending class in the morning may perform differently from students at the end of the day. Differences in the response variable may be due to the time of day and not to differences in the teaching methods. Thus the time of day would be a confounding variable. Choice (B) is another way in which this study is flawed. There are only two experimental units: the two classes. One is getting the lecture and the other is getting lecture combined with activities. There is only one experimental unit in each group, and therefore there is insufficient replication. Choice (C) demonstrates still another reason why this design is flawed. If one of the classes were to perform better than the other, we would not be able to tell whether this difference was due to the different method of teaching or a differing level of ability between the two classes. Choice (E) states that since he is using only statistics classes taught by the same teacher, he cannot generalize the results of the study to all classes in the school. If his original intent was, as stated in the problem, to determine if method of instruction affects how classes learn, the researcher might include other types of class than just statistics. Even if all other flaws in the experiment were addressed, applying the study only to statistics classes limits the applicability of the results to statistics classes.

33. **(D)** Coins have no memories. The first 7,000 tosses could have been all heads. That event is highly unlikely. The probability that the next toss is a tail is still ½. This question is <u>not</u> asking for the probability that, if a coin is tossed 6 times, the first 5 tosses are heads and the sixth one is a tail.

34. **(C)** The question asks which of the following conclusions can NOT be drawn from the given information. Choice (E) is a correct conclusion because since $P(B) = \frac{1}{2}$ and $P(B \text{ given } A) = \frac{1}{2}$, we know that A and B are independent. This is the definition of independent events. Since the events are independent, we know that $P(A \cap B) = P(A) \cdot P(B) = \frac{1}{2} \cdot \frac{3}{4} = \frac{3}{8}$. This makes choice (A) a true statement. We also know that $P(A \text{ or } B) = P(A) + P(B) - P(A \cap B) = \frac{1}{2} + \frac{3}{4} - \frac{3}{8} = \frac{7}{8}$. So, choice (B) is also a true statement. $P(A \text{ given } B) = P(A) = \frac{3}{4}$ because A and B are independent. Therefore choice (C) is not true and is the correct answer.

35. **(E)** Because the field has definitive differences in fertility and thus differences in yield the closer the plots are to the river, the sampling procedure should employ a method which stratifies by distance from the river. Each north/south column of ten 1-acre plots is a logical stratum in that the ten 1-acre plots on the western most end of the field has the largest yield and the ten 1-acre plots on the eastern end have the smallest yields. Like blocking for experimental design, stratifying for sampling requires creating groups which are as homogenous as possible. Each north/south column is similar in fertility. Randomly selecting 1 plot from each of the ten 1-acre plots in each of the ten north/south columns will reduce differences in yield due to other unknown factors that may be present in each north/south column. Choice (D) is an example of cluster sampling. The stratified random sampling described in choice (E) is a better choice than the cluster sampling described in choice (D). The 10 plots in the cluster sampling are all the same distances from the north and south borders of the field. There may be differences in the soil or other factors affecting the growth of the corn due to distances from the northern and southern borders of the field. Random selection in the stratified random sampling described in choice (E) will select plots at varying distances from the northern and southern borders of the field.

36. **(E)** The least squares process minimizes the sum of the squares of the error terms. Since the observed value is denoted by y_i and the predicted value by $\hat{y}_i$, the difference between the observed and the predicted denotes the error term $(y_i - \hat{y}_i)$. The square of the error term is $(y_i - \hat{y}_i)^2$, and the sum of the squares of the error terms is denoted by: $\sum(y_i - \hat{y}_i)^2$.

37. **(B)** Use the formula: $z = \dfrac{\hat{p} - p_o}{\sqrt{\dfrac{p_o(1 - p_o)}{n}}}$. In this case $p_o = 0.5$ because we are looking to determine if a majority, meaning greater than 50%, will favor a change in food selections in the cafeteria. The value for n is 100 because we sampled 100 students.

(The value of 4,000 would only be helpful if we were checking the assumption of independence. To check independence we would determine if the population size is at least ten times the sample size.) The value of $\hat{p}$ can be calculated as $\hat{p} = \frac{65}{100}$. Filling in the formula gives us: $z = \frac{0.65 - 0.5}{\sqrt{\frac{0.5(0.5)}{100}}}$.

38. **(C)** The power of the test can be calculated by taking $1 - \beta$ or one minus the Type II error probability. There is a "seesaw" relationship between α and β. For a given value of the parameter, the probability of a Type II error decreases as the value of α increases. Since power is $1 - \beta$, it would make sense that the smaller β is, the larger the power. Logically we can conclude that the larger α is, the smaller β is, and the larger the power of the test is. For an example, suppose we are using a test of $H_0: \mu = 0$ against a one sided alternative of $H_a: \mu > 0$. Suppose $\alpha = .05$. Using the first set of diagrams below, the curve on the left is the distribution of the sample mean if the null hypothesis is true. In the first diagram, the shaded area is the area under the curve representing α or the area for which we will reject the null hypothesis. To calculate power or Type II error probability, we need a given value of the parameter to be true. Let us (arbitrarily) say the real mean is $\mu = 3$. The second curve represents the true distribution of the sample mean. The power of the test is the probability of correctly rejecting the null hypothesis for a given value of the parameter. In the second set of curves, the black and gray area together is the power of the test, the probability of rejecting the null hypothesis if $\mu = 3$.

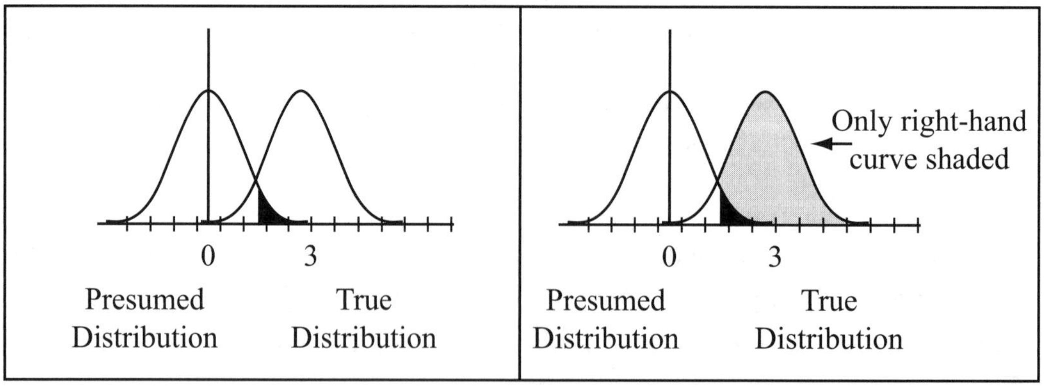

Performing the same test, but at a .10 level of significance, the shaded area under the curve in the first set of curves increases, and the second set of curves show that power increases.

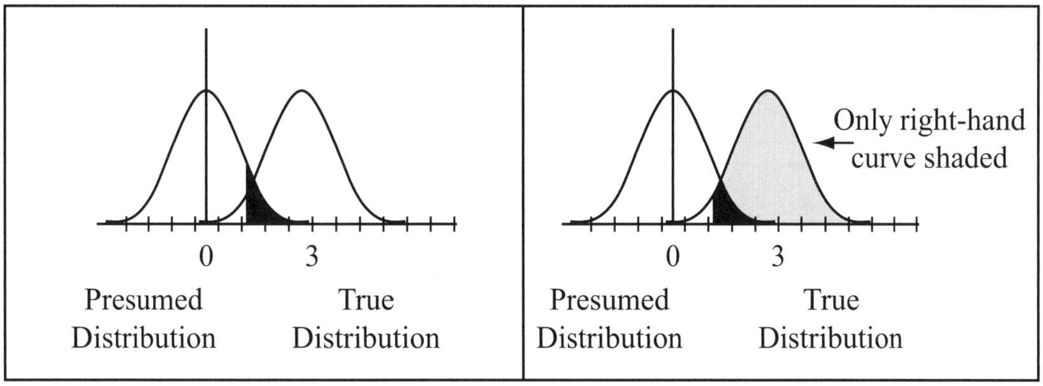

This also explains why choices (A) and (E) are incorrect. Choice (B) is incorrect because β decreases as sample size increases since the variability of the sampling distribution of the sample mean decreases as sample size increases. Using the same example as above, suppose we are using a test of $H_0: \mu = 0$ against a one-sided alternative of $H_a: \mu > 0$. Again suppose $\alpha = .05$. Using the first set of diagrams below, the curve on the left is the distribution if the null hypothesis is true. In the first diagram, the shaded area is the area under the curve representing α or the area for which we will reject the null hypothesis. In the second set of curves the black and gray area together is the power of the test, the probability of rejecting the null hypothesis when $\mu = 3$.

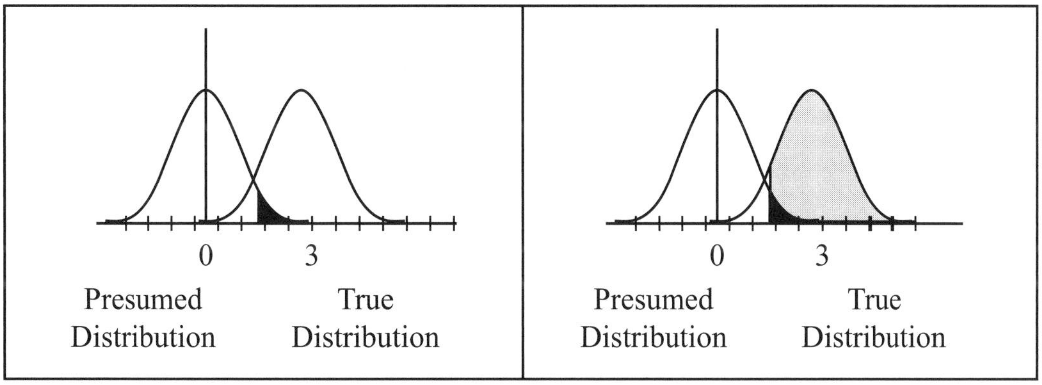

The two pictures below are the same as the ones above, except that the sample size increased, reducing the variability of the sample mean. The gray and black area combined is the power of the test. If the sample size increases, the power of the test is increased.

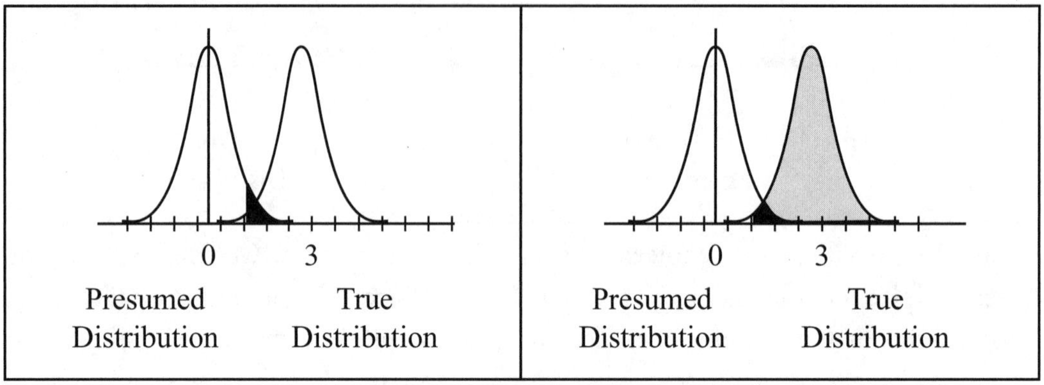

39. **(E)** Find b by substituting 2 for x and 5 for $\hat{y}$ into the equation $\hat{y} = 3x + b$ and solving for b. This will give you $b = -1$. You also know that $\overline{y} = a + b\overline{x}$. That gives us $\overline{y} = 3\overline{x} - 1$.

40. **(C)** The variance of the transformed data in degrees Celsius is equals $\left(\frac{9}{5}s_c\right)^2 = \frac{81}{25}s_c^2$, where s_c is the standard deviation of temperature in degrees Celsius.

Section II

Question One

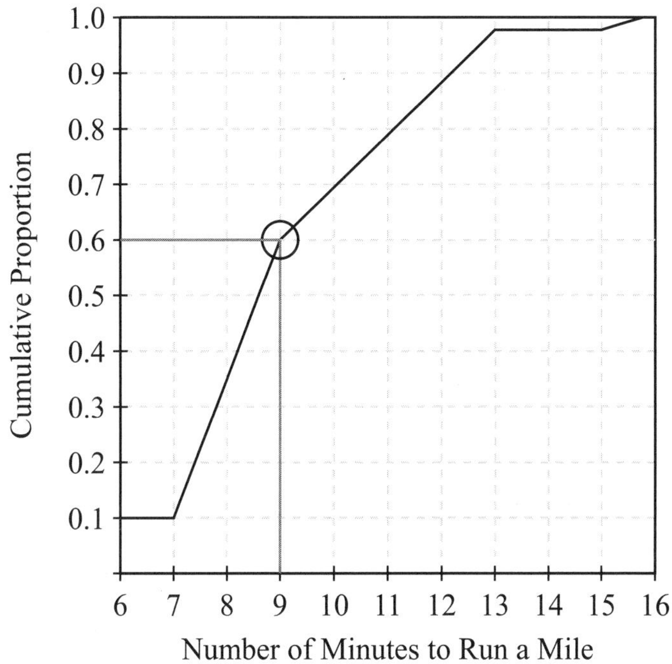

(a) The circled point indicates that 60% of the students ran the mile in 9 minutes or less. Each point on a cumulative frequency or relative frequency plot shows the proportion of values less than or equal to the specified value.

(b)

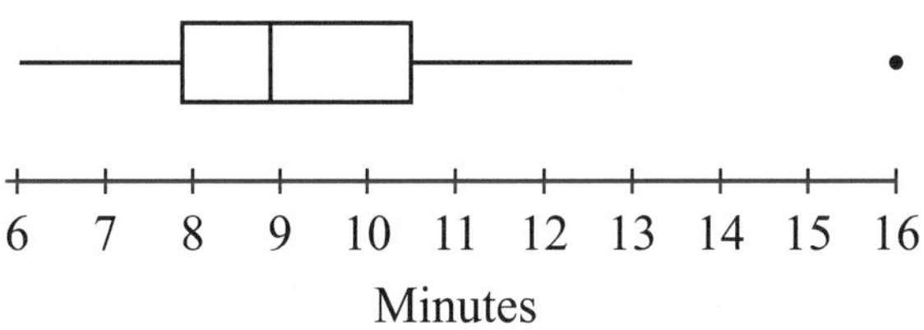

The minimum value is 6 minutes, the first quartile is about 7.7 minutes, the median is about 8.7 minutes, the third quartile is about 10.5 minutes, and the maximum value is 16 minutes. To find the first quartile, find 0.25 on the vertical axis and find the corresponding number of minutes on the horizontal axis. To find the median, look for the number of minutes corresponding to the 50th percentile. To find the third quartile, find 0.75 on the vertical axis and the corresponding value on the horizontal axis as shown below:

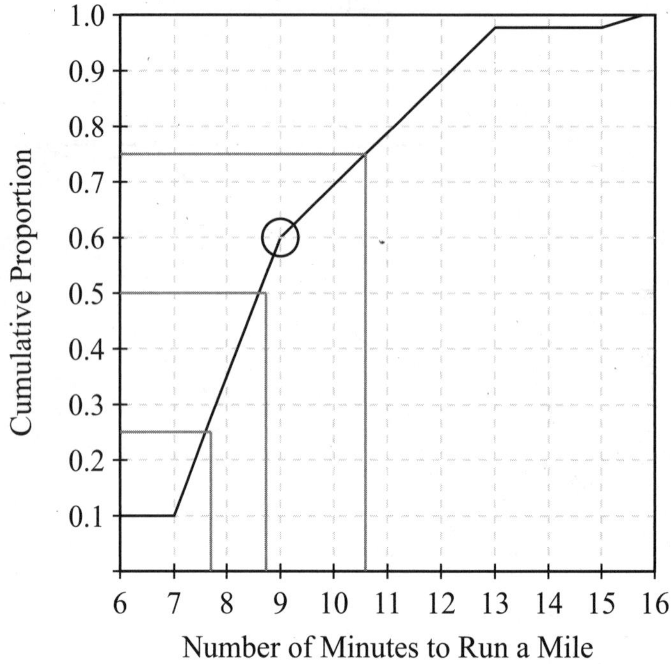

Next we need to check for outliers. Find the interquartile range or IQR. This can be calculated by taking Q3 – Q1. The IQR is 10.6 – 7.8 or about 2.8. The upper and lower fences for outliers are found by taking Q3 + 1.5(IQR) and Q1 – 1.5(IQR). These values are $10.5 + 1.5(2.8) = 14.7$ and $7.7 - 1.5(2.8) = 3.5$. Since the minimum value is 6, there are no outliers at the lower end. The maximum value of 16 is above the upper fence of 14.7 and is an outlier. Note that 97.5% of the 40 randomly sampled students ran the mile in 13 minutes or less. That number does not change for the values of 14 and 15 minutes. This indicates that no student ran the mile in 14 or 15 minutes and 2.5% of the 40 students or one student ran, the mile in 16 minutes. Thirteen minutes is the maximum number of minutes in which a student could run a mile that is not an outlier.

(c) The distribution is slightly right skewed with one outlier. (The outlier is not to be taken into consideration when describing the shape.) The center of the distribution is approximately 8.7 minutes (the median). Because the distribution is skewed, the interquartile range, about 2.8 minutes, is the best measure of the variability of the distribution.

(d) Because of the skewness of the distribution and the outlier above the upper fence, the mean is likely to be greater than the median of about 8.7. The mean is not a robust statistic and is pulled in the direction of the outlier or the skewness of the distribution. It is safe to assume that the mean of the sample will be greater than the average time of 8.5 minutes.

Question Two

(a) This is a completely randomized design where the unit of study is the smallest object to which a treatment will be assigned at random. Using this definition, it is the school that is the experimental unit.

(b) Schools vary in resources, quality of teachers, academic background of students, socioeconomic status of the communities that the schools serve, size of classes and number of students. All of these may affect the response variable, the differences in percentage of test scores that are proficient, and therefore are potential confounding variables. Wealthier districts generally fare better academically than do poorer districts. Smaller class size has a positive effect on student performance and standardized test results. Although we hope randomization of schools to treatments will reduce this problem as much as possible, we could have been so unlucky and ended up with all the poorer districts with large class sizes in one treatment group!

(c) To make all three groups as alike as possible, we could block by school. Randomly select 45 special education students from each of the 60 randomly selected districts to participate in the study. Randomly assign 15 students to each of the three treatments, class with one special education teacher, class with one special education teacher and one aide, and class with one special education teacher and one highly qualified mathematics teacher. By doing this we are controlling for class size (all three classes have 15 students) and ensuring that students in all three treatment groups in each school have the same socio-economic conditions.

Question Three

(a) $P(A = 0 \text{ and } B = 1) + P(A = 1 \text{ and } B = 2) + P(A = 2 \text{ and } B = 3) = 0.09 + 0.06 + 0.015 = 0.165$. The probability that squad A has one fewer call then squad B on any given day is the probability that the number of calls for squad A minus the number of calls for squad B equals negative one. There are three ways in which this can happen. Squad A can get zero calls and squad B can get one call, squad A can get one call and squad B can get two calls, or squad A can get two calls and squad B can get three calls. Since the numbers of calls received by the two squads are independent, we can multiply to get the probability of these events. *P*(squad A getting zero calls and squad B getting one call) $= (0.3)(0.3) = 0.09$. *P*(squad A getting one call and squad B getting two calls) $= (0.4)(0.15) = 0.06$. *P*(squad A getting two calls and squad B getting three calls) $= (0.15)(0.10) = 0.015$. Since these events are mutually exclusive, the sum of these probabilities is the probability that squad A has one fewer call than squad B. This is equal to $0.09 + 0.06 + 0.015 = 0.165$.

(b) The probability that squad A has fewer calls than squad B is the probability that the difference is either -1, -2, or -3. Since these are mutually exclusive events, the probability is $0.03 + 0.085 + 0.165 = 0.28$.

(c) $\mu_{X_A} = \sum(x_i \cdot p_i) = (0)(0.30) + (1)(0.40) + (2)(0.15) + (3)(0.10) + (4)(0.05) = 1.2$

$\sigma_{X_A} = \sqrt{\sum(x_i - \mu_X)^2 p_i}$

$\sigma_{X_A} = \sqrt{(0 - 1.2)^2(0.3) + (1 - 1.2)^2(0.40) + (2 - 1.2)^2(0.15) + (3 - 1.2)^2(0.10) + (4 - 1.2)^2(0.05)}$

$= \sqrt{1.26} = 1.1225$

(d) $\mu_A - \mu_B = 1.2 - 0.9 = 0.3$

$$\sigma_{\overline{X}_A - \overline{X}_B} = \sqrt{\frac{\sigma_A^2}{n_A} + \frac{\sigma_B^2}{n_B}} = \sqrt{\left(\frac{1.26}{36}\right) + \left(\frac{0.99}{36}\right)} = \sqrt{0.0625} = 0.25$$

(e) $P(\overline{X}_A - \overline{X}_B) < 0$ equals

$P\left(z < \frac{(\overline{X}_A - \overline{X}_B) - (\mu_A - \mu_B)}{\sigma_{\overline{X}_A - \overline{X}_B}}\right) = P\left(z < \frac{0 - 0.3}{0.25}\right) = P(z < -1.20)$ Using the standard normal table, we get the area to the left of -1.20 to be 0.1151.

Or, using the graphing calculator press 2nd distribution then scroll down to 2:normalcdf(-1E16, 0, 0.3, 0.25) and press enter.

Question Four

To determine whether the exercise program is effective in reducing weight a matched-pairs t procedure should be performed. A matched pairs procedure is used because the same people are weighed before the program and then again after the program.

1. Hypotheses:

$H_0: \mu_d = 0$

$H_a: \mu_d > 0$

Here μ_d is the mean difference between the pre-exercise weight and the post-exercise weight (pre-exercise weight − post exercise weight) in the population.

2. Name of test and conditions for inference:

Matched pairs t-procedure. Formula used is: $t = \dfrac{\overline{x}_d - 0}{\frac{s_d}{\sqrt{n}}}$. The mean of the difference between the pre-exercise weight and the post-exercise weight (pre-exercise weight – post exercise weight) in the sample is denoted by $\overline{x}_d$ and s_d is the standard deviation of these differences.

Conditions:

- The data are collected from a simple random sample. Stated in the stem of problem.
- The sample size is large or the distribution of the population of differences in weights is approximately normal.

To check the condition of normality, use a boxplot, histogram or a normal quantile plot on the difference (pre-exercise – post-exercise):

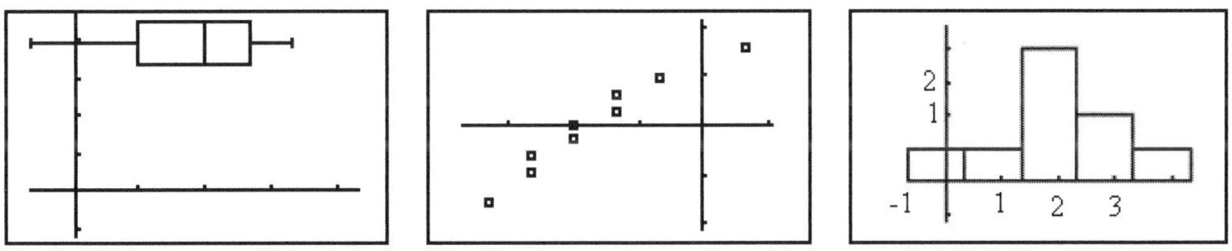

As the boxplot shows that the distribution is not strongly skewed and has no outliers, the normal quantile plot is roughly linear or the histogram is approximately mound shaped we can assume that the distribution of the population of differences is approximately normal. Since the sample data gives us no cause to question the assumptions regarding the normality of the population of differences, we assume that it is reasonable to use the t-test here. All of the above graphs can be constructed using the TI-84 calculator's STAT PLOT menu.

3. Mechanics:

Using formula: $t = \dfrac{\overline{x}_d - 0}{\frac{s_d}{\sqrt{n}}} = \dfrac{2.\overline{5}}{\frac{1.81}{\sqrt{9}}} = 4.24.$

To get the p-value, use the t distribution table and locate 8 degrees of freedom (number of differences − 1) in the first column. Look across until you find two numbers that bracket 4.23. These would be 3.833 and 4.501. Now look at the tail probability p for these two numbers. This indicates that the p-value is between .0025 and 0.001.

Tail probability	.25	.20	.15	.10	.05	.025	.02	0.1	.005	.0025	.001	.0005
df = 8	0.706	0.889	1.108	1.397	1.860	2.306	2.449	2.896	3.355	3.833	4.501	5.041

To get the p-value using the TI-84 calculator, press 2nd, VARS. Choose 6:tcdf(. Press ENTER. Fill in the parentheses with test statistic, a large number and the number of degrees of freedom: tcdf(4.24, E10, 8). This will give you a p-value of 0.0014.

4. Conclusion:
Since the p-value is 0.0014, which is less than 0.05, we reject the null hypothesis. We have convincing evidence that the weights of adults are, on average, greater before this program of exercise than after the program.

Question Five

(a) We are given that the percentage of students earning an A is 20% and that the percentage of students earning a B is 40%. By subtraction from 100% we can determine that the percentage of students earning a C or lower is 40%. Let D represent the event of getting a 2-year degree or better, let A represent the event of an A average in high school, let B represent the event of a grade average of B in high school, and let C represent the event of a grade of C or less in high school. The probability of a student who is known to have completed a 2- year degree or higher in fact having an average of A or B in high school is $P(A \mid D) + P(B \mid D)$. We know $P(A) = 0.20$, $P(B) = 0.40$, $P(C) = 0.40$, $P(D \mid A) = 0.64$, $P(D \mid B) = 0.37$, and $P(D \mid C) = 0.14$. We use the tree diagram below to compute *P*(*D*), the probability of getting a two-year degree or better.

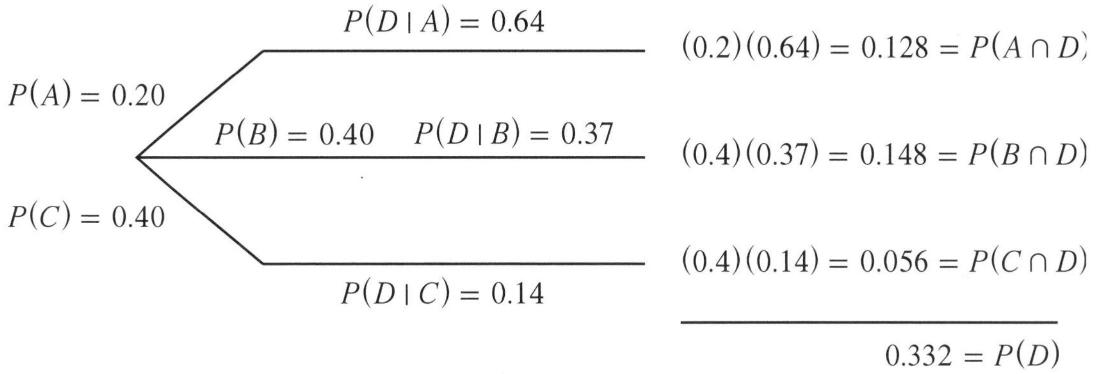

$$P(D) = P(A) \cdot P(D \mid A) + P(B) \cdot P(D \mid B) + P(C) \cdot P(D \mid C) = 0.332$$

We need $P(A \mid D)$ and $P(B \mid D)$.

$$P(A \mid D) = \frac{P(A \cap D)}{P(D)} = \frac{0.128}{0.332} = 0.3855$$

$$P(B \mid D) = \frac{P(B \cap D)}{P(D)} = \frac{0.148}{0.332} = 0.4458$$

The probability of getting an A or a B given that a 2-year degree or better was earned is $0.3855 + 0.4458 = 0.8313$ or about 83%.

(b) This question is asking us to perform a Chi-Square test of independence. Is completion of a college degree independent of grade earned in high school?

1. Hypotheses:

H_0: Successful completion of a college degree is independent of grades earned in high school.

H_a: Successful completion of a college degree is dependent on grades earned in high school.

2. Name of test and conditions:
This is a Chi-Square test of independence.

$$\chi^2 = \sum \frac{(\text{observed} - \text{expected})^2}{\text{expected}}$$

It is stated in stem of problem that the assumptions for inference have been met.

3. Mechanics:
First set up a table of count data as shown below and find row and column totals:

	Average High School Grades			
Subject	A	B	C or lower	Totals
Percentage Attaining 2-year degree or higher	$(0.64)(200) = 128$	$(0.38)(400) = 148$	$(0.14)(400) = 56$	332
Percentage not attaining any degree	$(0.36)(200) = 72$	$(0.36)(400) = 252$	$(0.86)(400) = 344$	668
Totals	$0.20(1000) = 200$	$0.40(1000) = 400$	$0.40(1000) = 400$	1000

In the null hypothesis, the assumption is that completion of a college degree and grades earned in college are independent. If that is true, then we would expect that the probability of completing a college degree and getting an A in high school, $P(A \cap D)$, would be equal to $P(A) \cdot P(D) = \left(\frac{200}{1000}\right) \cdot \left(\frac{332}{1000}\right) = 0.0664$. Therefore, the expected number of students in the sample who would earn a degree and have an A average in high school would be $(0.0664)(1000) = 66.4$. This process could be used to determine all of the expected counts. Alternatively, we could use the formula,

$$\text{Expected Count} = \frac{(\text{row total})(\text{column total})}{\text{grand total}}.$$

The observed and expected counts are shown in the table below.

Average High School Grades

Subject	A	B	C or lower	Totals
Percentage Attaining 2-year degree or higher	Observed = 128 Expected = $\frac{(332)(200)}{1000} = 66.4$	Observed = 148 Expected = $\frac{(332)(400)}{1000} = 132.8$	Observed = 56 Expected = $\frac{(332)(400)}{1000} = 132.8$	332
Percentage not attaining any degree	Observed = 72 Expected = $\frac{(668)(200)}{1000} = 133.6$	Observed = 252 Expected = $\frac{(668)(400)}{1000} = 267.2$	Observed = 344 Expected = $\frac{(668)(400)}{1000} = 267.2$	668
Totals	$0.20(1000) = 200$	$0.40(1000) = 400$	$0.40(1000) = 400$	1000

Now calculate X^2:

$$X^2 = \frac{(128-66.4)^2}{66.4} + \frac{(148-132.8)^2}{132.8} + \frac{(56-132.8)^2}{132.8} + \frac{(72-133.6)^2}{133.6} + \frac{(252-267.2)^2}{267.2} + \frac{(344-267.2)^2}{267.2}$$
$$= 154.64$$

The p-value can be found by using the Chi-Square distribution of Critical Values table with 2 degrees of freedom. The number of degrees of freedom is calculated using the formula, df = (row – 1)(column – 1). There are three columns and two rows in the table above. This gives us $(3-1)(2-1) = 2$ degrees of freedom. Since the test statistic, 154.64, is larger than 15.20, our p-value must be smaller than 0.0005.

4. <u>Conclusion</u>:
Since the p-value is so small, we must reject the null hypothesis. We have convincing evidence that there is an association between the grade earned in high school and completion (or not) of a college degree.

Section II
Part B

Question Six

(a) Side by side boxplots are shown below:

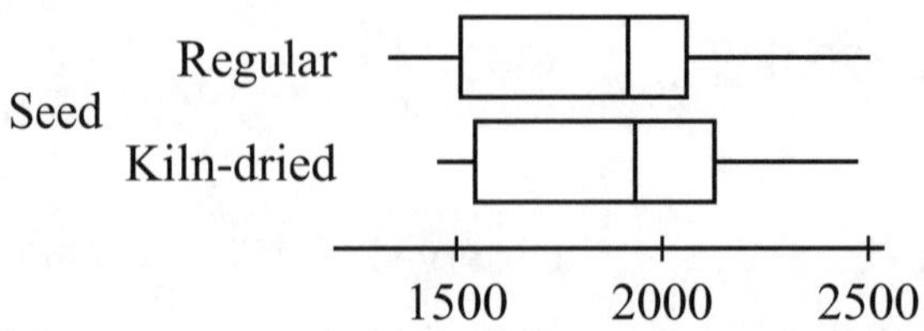

The two distributions are very similar. The centers and IQR's are about the same. Both are similar in shape. The range of yield of "kiln-dried" seed is slightly less than the range of the "regular" seed.

(b) In fitting a line to predict the yield of "kiln-dried" seed using yield of "regular" seed, we find that there is a very strong correlation between the two variables. The r value of 0.98 indicates a strong positive linear relationship between yield of "regular" seed and yield of "kiln-dried" seed. $r = \sqrt{r^2} = \sqrt{0.963} = 0.98$.

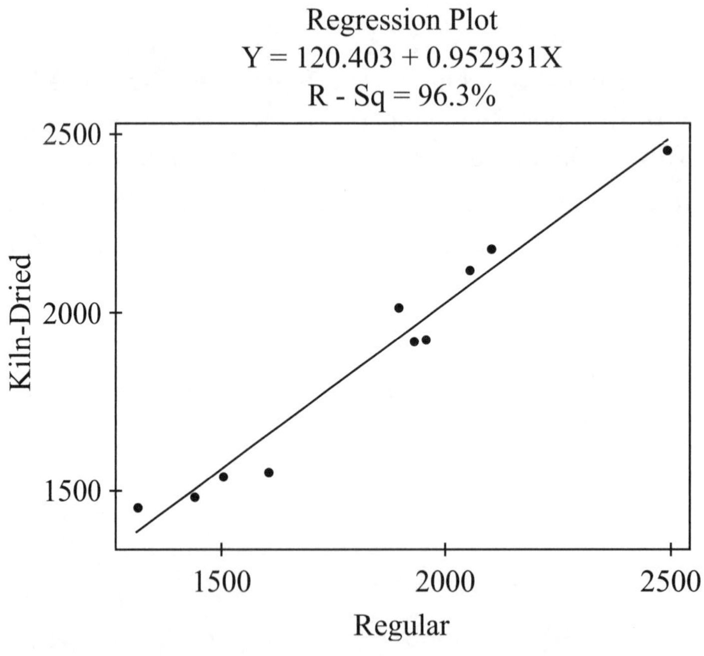

The scatterplot appears to be approximately linear indicating a line is appropriate to use to predict yield of "kiln dried seed" from the yield of "regular seed".

To test to see if the linear relationship is statistically significant, a test on the slope is carried out.

1. Hypotheses:

$H_0: \beta = 0$

$H_a: \beta \neq 0$

OR

H_0: There is no linear relationship between yield of "regular seed" and yield of "kiln-dried" seed.

H_a: There is a linear relationship between yield of "regular seed" and yield of "kiln-dried" seed.

2. Name of test and conditions:

Name of test: Linear regression *t*-test Formula: $t = \frac{b}{SE_b}$

In order to proceed with inference procedures for the slope, certain requirements must be met. They are:

- There is a random pattern in the residuals, indicating that a linear model is most appropriate.
- There must be no indication of an increase or decrease in the sizes of the residuals as the value of the explanatory variable increases.
- We should be confident that, in the population, the vertical deviations from the regression line are normally distributed

In testing the first requirement, look at the residual plot. The residual plot shows no pattern. Note: if a student claims a pattern, he/she should have clearly marked the pattern on the plot and stated concern about proceeding.

We can test the second requirement by again looking at the residual plot. Residuals with non-constant variance in the error terms would show a fan shape pattern in the residual plot. Since our plot does not show this type of pattern, we can safely assume that our residuals have a constant standard deviation.

We can test the last requirement by examining the normal quantile plot of the residuals and looking for signs of strong non-normality. There is some evidence of skewness, but the graph is fairly linear indicating that it is safe to assume that, in the population, the vertical deviations from the regression line are normally distributed.

3. Mechanics:

To obtain the test statistic we use the formula: $t = \frac{b}{SE_b} = \frac{.95293}{0.06241} = 15.27$.

The *p*-value for this test statistic is 0.000 (reading the computer output, "REGULAR" row under the "P" column) or $9.67(10^{-8})$ using the TI-84 graphing calculator's LinReg *t*-test option.

4. Conclusion:

Since the *p*-value is small, we reject the null hypothesis. We have convincing evidence of a linear relationship between the yields using kiln dried seed and regular seed. Therefore we are justified in using the linear regression model to predict the yield using kiln dried seed from the yield using regular seed.

(c) This question is asking if there is a difference in average yield for the two types of seed. It is a paired t-test because each plot was planted with both types of seed.

1. Hypotheses:

   H_0: $\mu_d = 0$ The population mean of the differences is 0.

   H_a: $\mu_d \neq 0$ The population mean of the differences is not 0.

2. Name of test and conditions:

   Name of test: paired t-test. Formula: $t = \dfrac{\overline{x}_d - \mu_d}{\frac{s_d}{\sqrt{n}}}$

   The condition is that the distribution of the population of differences be approximately normal. Using a boxplot of the differences, we get the boxplot shown below, which shows that the distribution is roughly symmetrical and has no outliers. There is no reason to believe the distribution of differences is not approximately normally distributed. We will proceed with the test.

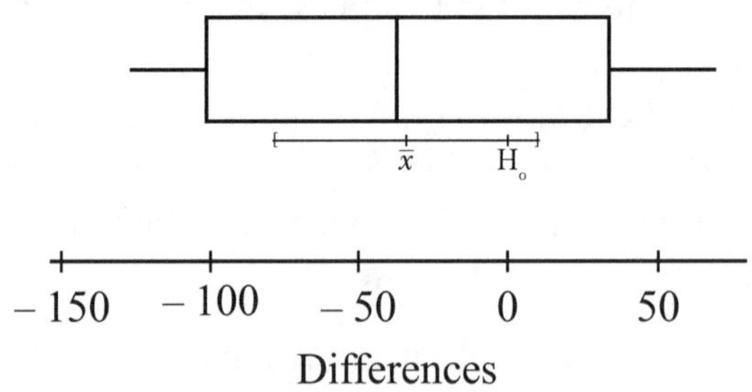

3. Mechanics:

   Compute the differences in yield for each plot of land and calculate the summary measures for these differences.

   Mean = – 33.7

   Standard Deviation = 66.2

   $n = 11$

   We use the formula: $t = \dfrac{\overline{x}_d}{s_d/\sqrt{n}} = \dfrac{-33.7}{66.2/\sqrt{11}} = -1.69$. The p-value can be found by looking up $t = 1.69$ with 10 degrees of freedom on the t-table. 1.69 is between 1.372 and 1.812 which gives us an upper-tail probability between 0.10 and 0.05. Doubling these values (since this is a two-tailed test), we find that the p-value is between 0.20 and 0.10. To obtain an exact p-value of 0.122 by means of the TI-84 calculator, use 1 – tcdf(-1.69, 1.69, 10).

4. Conclusion:

   Since our p-value of 0.122 is greater than 0.05, we do not reject the null hypothesis at the 0.05 level of significance. We do not have sufficient evidence to conclude that the mean yield of regular seed and the mean yield of kiln dried seed are different.

Sample Examination Four – Answers

1. **(B)** $(20/100) \cdot 24 = 4.8$. So the 20th percentile travel time is approximately the 5th time, counting from the smallest to the largest. So the 20th percentile time is approximately 40 minutes.

2. **(B)** We need the probability of success to be 1/3. In the strategy in (B), three digits represent success and six digits represent failure, and so this gives a probability of success that is equal to 1/3, as is required; thus, (B) is the correct response.

   In the strategy in (A), three digits represent success and seven digits represent failure. This gives a probability of success that is equal to 3/10, which is incorrect. In the strategy in (C), 33 digit-pairs represent success and 67 digit-pairs represent failure. This gives a probability of success that is equal to $33/100 = 0.33$, which is incorrect. In the strategy in (D), 30 digit-pairs represent success and 70 digit-pairs represent failure. This gives a probability of success that is equal to $30/100 = 0.3$, which is incorrect. In the strategy in (E), 334 three-digit numbers represent success and 666 three-digit numbers represent failure. This gives a probability of success that is equal to $334/1000 = 0.334$, which is incorrect.

3. **(E)** Whenever possible, it is preferable to perform a census (that is, gather information about the whole population) rather than take a sample. The strategy in (E) achieves this.

4. **(E)** The only way to establish causation is by performing an experiment. Randomly assigning some teenagers to play video games and some not to play video games constitutes an experiment; all the other options are modifications of the observational study described in the question.

5. **(E)** The quantity $36/50 = 0.72$ is the *sample* proportion, and so should not be used in the hypotheses. The alternative hypothesis should be $p > 0.65$, since the proportion of employees who eat the meals is suspected to have *increased* beyond the proportion of 0.65 that has been assumed.

6. **(A)** r^2 is the proportion of the variation in y that can be explained by the least squares regression line of y on x.

7. **(D)** The data given are *paired*, and so of the three options given, only the paired t-test and the t-test for the slope of the regression line can be applied. The two-sample t-test is used when you have two independent random samples, which is not the case here.

8. **(D)** The number of sixes rolled has a binomial distribution with $n = 5$ and $p = 1/6$. So, denoting the number of sixes rolled by X, $P(X = 1) = {}_5C_1(1/6)(5/6)^4 = 0.4019$, and $P(X = 2) = {}_5C_2(1/6)^2(5/6)^3 = 0.1608$. Therefore, $P(X = 1 \text{ or } X = 2) = 0.4019 + 0.1608 = 0.5627$.

9. **(E)** In order to use the t distribution we have to know either that the population is normally distributed or that the sample is large (usually requiring $n > 30$). Here $n = 15$, and so we examine the distribution of the sample values in order to check that it is feasible that the population is normally distributed.

10. **(C)** It is clear from the boxplot that the distribution is positively skewed. Hence the mean is greater than the median. It can probably be seen intuitively that the mean cannot be as large as 22, but to prove this, consider the fact that the five number summary (read from the boxplot) is 0, 5, 10, 20, 30. Since there are 23 measurements, this tells us that the 1st, 6th, 12th, 18th, and 23rd values are 0, 5, 10, 20, and 30, respectively. In other words, the dataset, in order from the smallest to the largest, is:

    $\underline{0}$ * * * * $\underline{5}$ * * * * * $\underline{10}$ * * * * * $\underline{20}$ * * * * $\underline{30}$,

    where the stars represent any numbers, so long as the list remains in numerical order. Therefore the largest possible value of the mean is achieved by the following dataset:

    $\underline{0}$ 5 5 5 5 $\underline{5}$ 10 10 10 10 10 $\underline{10}$ 20 20 20 20 20 $\underline{20}$ 30 30 30 30 $\underline{30}$.

    The mean of this set of values is 15.43, and so this is the largest possible mean for a dataset of size 23 having the boxplot given in the question.

11. **(D)** The confidence interval for the difference of two proportions is

    $$(\hat{p}_1 - \hat{p}_2) \pm z^* \sqrt{\frac{\hat{p}_1(1-\hat{p}_1)}{n_1} + \frac{\hat{p}_2(1-\hat{p}_2)}{n_2}}.$$

    The value of z^* is found using the calculator by means of invNorm(.995), or the equivalent. (On the TI-83 and 84 calculators, "invNorm(" is accessed by selecting 2nd,DISTR,3.) This gives $z^* = 2.576$. Therefore, since $\hat{p}_1 = 83/120$, $\hat{p}_2 = 78/140$, $n_1 = 120$ and $n_2 = 140$ the interval given in (D) is correct.

    Note that (C) and (E) make use of the proportion of germinating seeds for the *combined* samples. This is not valid in the context of confidence intervals.

12. **(E)** We are testing whether the three category proportions are the same for the two populations. This is a chi-square test for homogeneity.

    Note that the tests given in (A) and (D) are applicable to studies involving a *single* sample. This study involves two samples, and therefore these responses are incorrect. Options (B) and (C) are tests for means. This study concerns proportions, and so these responses are incorrect.

13. **(D)** Replication is exactly as described in option (D): using enough subjects so that chance variations between the groups become negligible.

14. **(E)** The distribution with the smallest standard deviation is the one with the greatest concentration close to its mean. This criterion is satisfied by the distribution represented by the histogram in (E), since the great majority of the data values are at most one unit from the mean. All of the other options show deviations from the mean that are, on average, greater than those in (E).

15. **(C)**

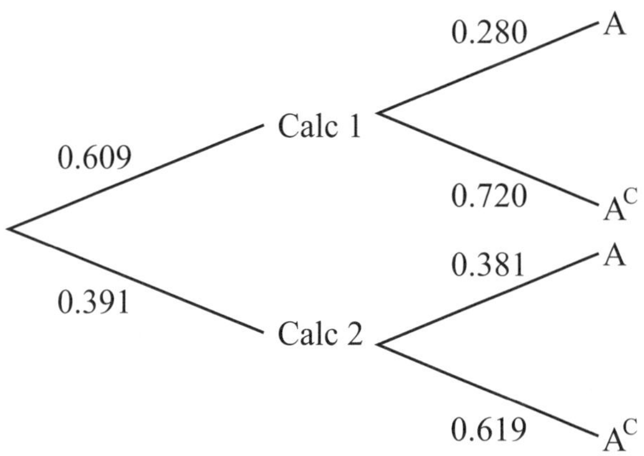

$$P(\text{Calc } 1 \mid \text{A}) = \frac{P(\text{Calc } 1 \cap \text{A})}{P(\text{A})} = \frac{0.609 \cdot 0.28}{0.609 \cdot 0.28 + 0.391 \cdot 0.381} = 0.534.$$

16. **(D)** The computer output gives "Prob < t" as 0.0297: this is the *p*-value for the one-tailed test specified in the question. It is also clear from the output that the students' scores have decreased on average, since the PostTest mean is less than the PreTest mean. So we have evidence at the 5% level, but not the 1% level, that the students' mental arithmetic abilities have decreased. Thus, option (D) is correct (and option (E) is incorrect).

Options (A) and (B) are incorrect since the PostTest mean is less than the PreTest mean, and so we cannot possibly have evidence that on average the students' mental arithmetic scores have increased. Turning our attention to (C), we note that this option refers to the equivalent *two*-tailed test. The *p*-value for the one-tailed test is 0.0297, and so the *p*-value for the two-tailed test would be $2(0.0297) = 0.0594$. Hence we would not have evidence at the 5% level of a change in the students' average score; thus, option (C) is incorrect.

17. **(B)** $\frac{\sigma}{\sqrt{n}}$ is the standard deviation of the distribution of the sample *mean*, not the standard deviation of the sample.

    Let us now look at why the statements given in the other options are correct (making these incorrect responses). The distribution of $\bar{x}$, the sample mean, has mean μ and standard deviation $\sigma/\sqrt{n}$; thus, the statements in (A), (D), and (E) are correct. Moreover, the Central Limit Theorem tells us that for large values of *n*, the distribution of $\bar{x}$ is approximately normal. Hence the statement in (C) is correct.

18. **(A)** To say that the distribution is negatively skewed means that it has a longer tail to the left than to the right. Option (A) is the only response that satisfies this. Note that (D) and (E) are incorrect as they tell us nothing about the *spread* of relatively high scores, as compared to the *spread* of relatively low scores. Note, also, that (C) is incorrect, since the histogram of a negatively skewed distribution will have its highest point towards the right, not the left.

19. **(A)** There is no reason to conclude that the study was conducted poorly. Type I and Type II errors occur by *chance*. Typically, when a Type I or II error occurs, the survey or experiment is well designed and conducted, and the results are correctly analyzed, however it happens by chance that the results lead to the correct conclusion not being reached by the significance test.

    Let us now look at why the statements in (B)-(E) are correct (making these responses incorrect). A Type II error is failing to reject H_0 when H_0 is false, and this is failure to reach a correct conclusion; thus, the statements in (B), (C) and (D) are correct. Turning our attention to (E), we note that a Type I error is rejecting H_0 when H_0 is *true*. We are told here that a Type II error has occurred, and therefore we know that H_0 is *false*. Therefore, a Type I error cannot have occurred, and the statement in (E) is true.

20. **(B)** The difference between experiments and observational studies is that in experiments the conditions are *imposed* on the participants, whereas in observational studies the participants have made their own choices as to what they do. Thus, (B) is the correct response.

21. **(A)**

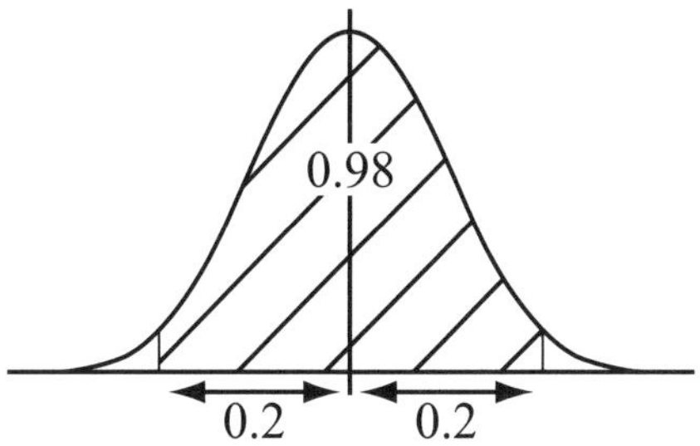

The areas of the left and right tails in the diagram are 0.01. So the z-value for the upper bound of the shaded region in the diagram is given by invNorm(.99), or the equivalent. (On the TI-83 and 84 calculators, "invNorm(" is accessed by selecting 2nd,DISTR,3.) This gives $z = 2.326$. The quantity z is the number of standard deviations from the mean. Hence $0.2 = 2.326\sigma$, which gives $\sigma = 0.086$.

22. **(C)** $t = \dfrac{b - \beta}{s_b} = \dfrac{4.169 - 0}{2.142} = 1.946$.

23. **(B)** Bias lies in the sampling *method.* Bias is present when something in the sampling *design* makes it likely that the sample will over- or under-emphasize some relevant characteristic of the population. Thus, (B) is the correct response. The phenomenon in (A) – getting a sample that over-emphasizes some characteristic of the population – can occur purely as a result of sampling variability, and is therefore not indicative of bias.

24. **(B)** The z-score for a particular value tells us how many standard deviations that value is above (or below) the mean, with negative z-scores signifying values below the mean. Therefore, with a z-score of -0.290, this item is 0.290 standard deviations below the mean. Hence the price of the item is $3.4573 - 0.290 \cdot 1.613 = 2.99$.

25. **(C)** The statement in (C) is a correct interpretation of the 95% confidence level. Looking at the other options, remember that any mention of the particular confidence interval that has been calculated – as in (A) and (B) – makes an attempted interpretation of the confidence level incorrect. Considering (D), note that the confidence interval that has been constructed gives us some information about the mean of the *population.* Any attempt to apply the confidence interval to a *sample* mean, as in (D), is incorrect. Finally, considering (E), note, again, that the confidence interval tells us something about the population *mean.* Any attempt to relate the interval to the actual population measurements without reference to the population mean, as in (E), is incorrect.

26. **(C)** Recall that when X and Y are independent random variables and a and b are constants, $\sigma_{aX+bY} = \sqrt{a^2\sigma_X^2 + b^2\sigma_Y^2}$. Here, if X is the amount (in grams) of chips and Y is the amount (in grams) of salsa, then the number of calories is $5X + 0.5Y$. Therefore the standard deviation of the number of calories is $\sqrt{5^2(3.7)^2 + 0.5^2(2.8)^2} = 18.55$.

27. **(A)** The total score for the girls is $12 \cdot 8.25 = 99$. The total score for the boys is $20 \cdot 7.3 = 146$. There are $12 + 20 = 32$ students altogether. Therefore, the mean score for all of the students is

$$\frac{99 + 146}{32} = 7.66.$$

28. **(C)** The probability that the largest score is a six is the probability that at least one six is rolled, and this probability is $1 - P(\text{no sixes}) = 1 - \left(\frac{5}{6}\right)^4 = 0.518$.

29. **(B)** Since 18 out of the 30 items of clothing in the sample have been worn, the sample proportion is $\hat{p} = 18/30 = 0.6$. Denoting the proportion of all returned items of clothing that have been worn by p, we need to test the null hypothesis $H_0: p = 0.5$ against the alternative hypothesis $H_a: p > 0.5$. We do this using a one-proportion z-test, for which the test statistic is

$$z = \frac{\hat{p} - p}{\sqrt{\frac{p(1-p)}{n}}} = \frac{0.6 - 0.5}{\sqrt{\frac{(0.5)(0.5)}{30}}} = 1.095.$$

Therefore the p-value for the test is given by $P(Z > 1.095) = 0.137$. (This is calculated using normalcdf(1.095,999), or the equivalent; on the TI-83 and 84 calculators, "normalcdf(" is accessed by selecting 2nd,DISTR,2.)

30. **(B)** $s = \sqrt{\frac{\sum(x - \bar{x})^2}{n - 1}} = \sqrt{\frac{3.24 + 3.24 + 17.64 + 0.04 + 0.64}{4}} = 2.49.$

31. **(B)** For 50 rolls of the die, the probability that the total score is ≥ 200 is the probability that $\overline{X} \geq 4$, where $\overline{X}$ is the mean score for the 50 rolls. Since $n = 50 > 30$, the distribution of $\overline{X}$ is approximately $N\left(\mu, \frac{\sigma}{\sqrt{n}}\right) = N\left(3.5, \frac{1.708}{\sqrt{50}}\right)$. So $P(\overline{X} \geq 4)$ is approximately the shaded area in the diagram below.

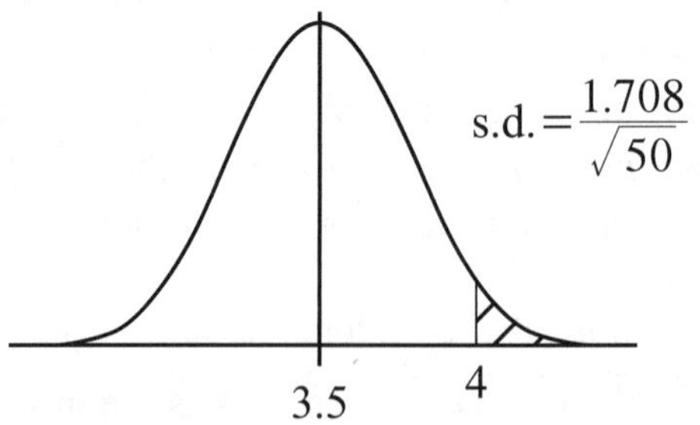

The z-value for the lower bound of the shaded region is $z = \dfrac{4 - 3.5}{\left(1.708/\sqrt{50}\right)} = 2.06998.$

So the approximate value of $P(\overline{X} \geq 40)$ is calculated by means of normalcdf(2.06998,999), or the equivalent, which gives 0.019 as the answer. (On the TI-83 and 84 calculators, "normalcdf(" is accessed by selecting 2nd,DISTR,2.)

32. **(C)** The 95% confidence interval is given by $\overline{x} \pm 1.96 \cdot \dfrac{\sigma}{\sqrt{n}}$. Therefore, when n is multiplied by 4, the width of the confidence interval is multiplied by 1/2.

33. **(A)** We will denote the y-intercept and the slope of the regression line by a and b, respectively. We are given $\overline{x} = 70.2$, $\overline{y} = 160.3, b = 2.75$. The formula sheet gives $b_0 = \overline{y} - b_1\overline{x}$, and so in our notation, $a = \overline{y} - b\overline{x} = 160.3 - 2.75 \cdot 70.2 = -32.75$. Hence the equation of the regression line is $\hat{y} = -32.75 + 2.75x$. Substituting $x = 68$, we get $\hat{y} = -32.75 + 2.75 \cdot 68 = 154.25$.

34. **(B)** As indicated in the hypotheses, the hypothesis test for the slope of the regression line is also a test for correlation. The computer output gives the p-value for this test as 0.094, which is the p-value for the two-tailed test. (The p-value given by a computer can be assumed to be the one for a two-tailed test unless it is specified as being for the one-tailed version.) Therefore, since $0.094 > 0.05$, we do not have sufficient evidence at the 5% level to conclude that there is a nonzero correlation between the results of the two methods. Note that (C) is "accepting H_0," which is never correct.

35. **(A)** The probability density function of the chi-square distribution with 8 degrees of freedom is shown in the diagram below.

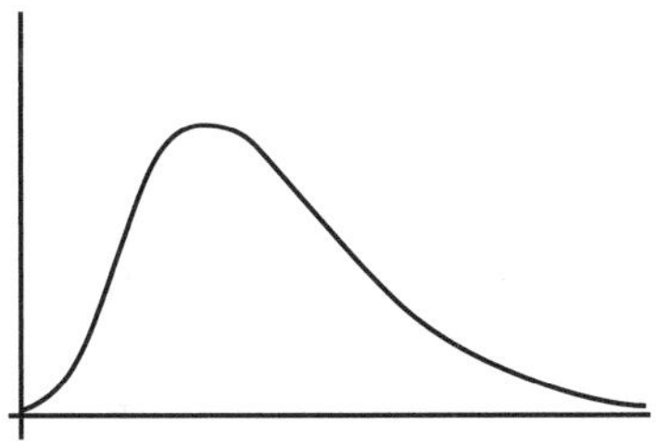

This distribution is not symmetrical. The distributions given in options (B)-(E) are all symmetrical.

36. **(C)** This question is about the placebo effect. The point of any experiment is to keep all variables equal (or as close as possible) between the two groups except for the one being tested, which here is the chemical in the drug. But in the design given, there is an additional difference between the two groups in that the members of Group A are taking a tablet whereas those in Group B are not – and we know that the mere taking of a tablet can have a positive effect. Hence (C) provides a strong argument against the claim.

Option (A) is incorrect since the mathematics involved in significance testing always takes into account the sizes of the groups/samples. Looking at (B), we have no reason to think that the people in Group A might be eating less than the people in Group B since the groups are chosen at random, unless the taking of the tablets is causing a reduction in eating, which would in fact be a weight-reducing effect of the tablets. The warmer weather cited in (D) would affect the experimental subjects in both groups, and so cannot account for a greater weight loss on average in Group A. Finally, considering (E), if some of the people in Group A had not taken the drug (which in fact constitutes a breakdown in the experimental procedure), then the observed weight loss for Group A would most likely have been caused by a very significant weight loss in those in Group A who *did* take the drug. This would *reinforce* the evidence for the effectiveness of the tablets.

37. **(B)** First recall, for example, that zero being contained in the 95% confidence interval is equivalent to the non-rejection of H_0 at the 5% level for the two-tailed test. Here the p-value for the *one*-tailed test is given as 0.043, and so the p-value for the two-tailed test is 0.086. Therefore H_0 is rejected at the 10% level, but is not rejected at the 8% level. Equivalently, zero is not contained in the 90% confidence interval but *is* contained in the 92% confidence interval. Thus, (B) is the correct response.

38. **(D)** The computer output gives the equation of the regression line to be $\hat{y} = 602.509 + 24.927x$, where x is the size and $\hat{y}$ is the predicted rent. So for $x = 32$ we get $\hat{y} = 602.509 + 24.927 \cdot 32 = 1400.173$. Therefore, for the point $(32, 1450)$ the residual is $y - \hat{y} = 1450 - 1400.173 = 49.827$. Option (D) gives the closest value to this result.

39. **(B)**

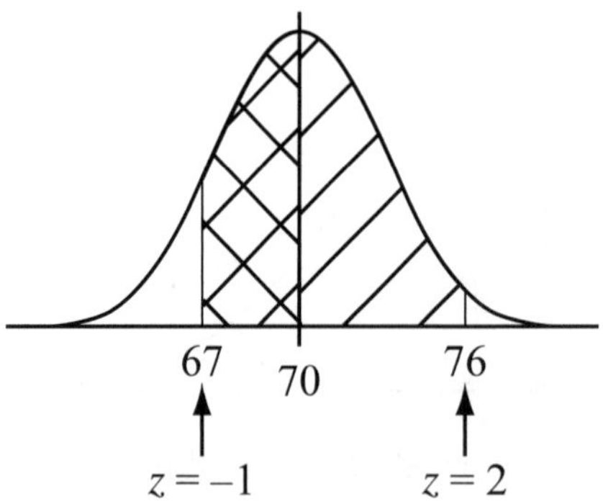

We require $P(X < 70 \mid 67 < X < 76) = \dfrac{P(X < 70 \text{ and } 67 < X < 76)}{P(67 < X < 76)}$.

Now, saying that $X < 70$ and $67 < X < 76$ is equivalent to saying that $67 < X < 70$. So we require

$$\frac{P(67 < X < 70)}{P(67 < X < 76)},$$

which is given by normalcdf(−1,0)/normalcdf(−1,2), or the equivalent. This calculation gives the answer: 0.417. (On the TI-83 and 84 calculators, "normalcdf(" is accessed by selecting 2nd,DISTR,2.)

40. **(A)** Consider the chi-square goodness of fit test of H_0: the four outcomes on the spinner are equally likely *versus* H_a: the four outcomes on the spinner are not equally likely. The quantities f_1, f_2, f_3, and f_4 are the observed frequencies of the four outcomes and, assuming H_0 to be true, the expected frequencies are all 30. Therefore, the quantity

$$Q = \frac{(f_1 - 30)^2}{30} + \frac{(f_2 - 30)^2}{30} + \frac{(f_3 - 30)^2}{30} + \frac{(f_4 - 30)^2}{30}$$

is the chi-square statistic $\sum \frac{(O - E)^2}{E}$, where O represents the observed frequency and E represents the expected frequency.

We know H_0 is true, since the question states that the four outcomes are equally likely. Also, the conditions for the chi-square goodness of fit test are satisfied since the expected frequencies are all greater than 5. So the distribution of Q is chi-square with 3 degrees of freedom (since there are 4 categories in this goodness of fit test).

So the required probability is $P(\chi_3^2 > 6.25) = 0.1001$. (This is calculated using χ²cdf(6.25,999,3), or the equivalent; on the TI-83 and 84 calculators, "χ²cdf(" is accessed by selecting 2nd,DISTR,7.) Thus, the correct response is (A).

Section II

Question One

(a) Approximately 91% of the country's residents are less than or equal to 70 years old.

(b) Looking at the graph, the cumulative relative frequency for Age $=40$ is approximately 0.58. This tells us that the proportion of residents over 40 years old is approximately $1 - 0.58 = 0.42$. Therefore the number of residents over 40 years old is approximately $32465233 \cdot (0.42) \approx \boxed{13.6 \text{ million}}$.

[*Note that you do not have to get this answer, exactly. So long as you have shown your work on the graph and your calculations are consistent with the work shown you will receive full credit. This applies to parts (a)-(c) of this question.*]

(c) Using the graph, the age for cumulative relative frequency $=0.75$ is approximately 52, and the age for cumulative relative frequency $=0.25$ is approximately 17.5. So the interquartile range is approximately $52 - 17.5 = \boxed{34.5 \text{ years}}$.

(d) The fact that the graph is approximately a straight line for 0-50 on the horizontal axis tells us that the numbers of residents in the age groups 0-5, 5-10, 10-15, and so on up to 45-50, are all approximately equal.

[*To understand this, note first that the vertical increment between, for example, Age = 10 and Age = 15 on the cumulative relative frequency graph is the proportion of residents who are between 10 and 15 years old. The fact that the graph is approximately a straight line tells us that these vertical increments are approximately equal, and therefore that the proportions of residents in the age groups 0-5, 5-10, and so on up to 45-50, are equal. It follows that the numbers of residents in these age groups are equal.*]

Question Two

(a) The random pattern in the residual plot tells us that the use of a straight line to model the relationship between log(number of subscribers) and Years since 1920 is appropriate.

[*Note: Do not attempt to use the value of r^2 to answer this question. The value of r^2 (or of r) can tell you how close the points are to being in a straight line, but it cannot tell you whether a straight line is an appropriate model.*]

(b) The equation of the regression line is:

$\text{Predicted }\log(\text{number of subscribers}) = 3.364928 + 0.0315783(\text{Yrs since }1920).$

[*Note that if you use symbols such as x and y in your equation, then you must say what these symbols represent.*]

(c) Substituting Yrs since 1920 $= 110$, we get

$\log(\text{number of subscribers}) = 3.364928 + 0.0315783(110) = 6.838541.$

Therefore, number of subscribers $= 10^{6.838541} = \boxed{6{,}895{,}107}$.

[*Note that different answers will be obtained according to the degree of rounding in the work. If your work is correct, and your answer agrees with your work, then you will receive full credit.*]

(d) The year 2030 lies outside the range of years in the original data set, and we have no reason to believe that the linear relationship between log(number of subscribers) and Yrs since 1920 will apply outside that range. Therefore, using this regression model to predict the number of subscribers in 2030, known as extrapolation, is not reasonable.

Question Three

(a) It could be argued that people who feel strongly against the increased police presence are more likely to respond than those who are neutral or in favor of the policy, in which case the survey would be likely to overestimate the level of opposition to the policy.

(b) The group could send the survey to a large random sample of residents. After a period of time, they could make follow-up phone calls to those residents who have not responded, in order to obtain responses from all of the residents in the sample.

(c) Let the proportion of residents in opposition to the policy be p. We estimate p using the sample proportion, $\hat{p}$. If the sample size, n, is large (we hope to get an answer that satisfies the condition that np and $n(1-p)$ are both greater than 10), then the distribution of $\hat{p}$ is approximately normal with mean p and standard deviation $\sqrt{\frac{p(1-p)}{n}}$. For any normally distributed random variable, with 95% probability the random variable will be within 1.96 standard deviations of the mean (we get 1.96 using invNorm(.975), or the equivalent; on the TI-83 and 84 calculators, "invNorm(" is accessed by selecting 2nd,DISTR,3). So, with 95% probability, $\hat{p}$ will be within $1.96 \cdot \sqrt{\frac{p(1-p)}{n}}$ of p.

Therefore, referring to the question, we need

$$1.96 \cdot \sqrt{\frac{p(1-p)}{n}} < 0.02.$$

From this we get

$$\sqrt{n} > \frac{1.96}{0.02}\sqrt{p(1-p)},$$

and so, squaring both sides, we see that we require

$$n > \left(\frac{1.96}{0.02}\right)^2 p(1-p).$$

We need a value of n that will satisfy this inequality for all values of p in the range $0.32 \le p \le 0.45$. Experimenting with different values of p, we see that $p(1-p)$ is largest when p is as close as possible to 0.5, in other words when $p = 0.45$.

So we need $n > \left(\frac{1.96}{0.02}\right)^2 \cdot 0.45 \cdot (1 - 0.45) = 2376.99$

So $n = \boxed{2377}$ is the sample size required.

Question Four

(a)

Sample	Sample Mean	Sample Midrange
1, 2, 2*, 3	2	2
1, 2, 2*, 3*	2	2
1, 2, 2*, 4	2.25	2.5
1, 2, 3, 3*	2.25	2
1, 2, 3, 4	2.5	2.5
1, 2, 3*, 4	2.5	2.5
1, 2*, 3, 3*	2.25	2
1, 2*, 3, 4	2.5	2.5
1, 2*, 3*, 4	2.5	2.5
1, 3, 3*, 4	2.75	2.5
2, 2*, 3, 3*	2.5	2.5
2, 2*, 3, 4	2.75	3
2, 2*, 3*, 4	2.75	3
2, 3, 3*, 4	3	3
2*, 3, 3*, 4	3	3

(b)

Sample Midrange	2	2.5	3
Probability	4/15	7/15	4/15

(c) $E(\text{sample mean}) = 2 \cdot \frac{2}{15} + 2.25 \cdot \frac{1}{5} + 2.5 \cdot \frac{1}{3} + 2.75 \cdot \frac{1}{5} + 3 \cdot \frac{2}{15} = \boxed{2.5}$.

$E(\text{sample midrange}) = 2 \cdot \frac{4}{15} + 2.5 \cdot \frac{7}{15} + 3 \cdot \frac{4}{15} = \boxed{2.5}$.

(d) The fact that the expected value of the sample mean is equal to the population mean tells us that use of the sample mean to estimate the population mean is appropriate, since, on average, the sample mean will be equal to the population mean.

The fact that the expected value of the sample midrange is equal to the population mean tells us that use of the sample midrange to estimate the population mean is also appropriate, since, on average, the sample midrange will be equal to the population mean.

(e) The standard deviation of a statistic is a typical deviation of the values of the statistic from the expected value of the statistic. Since the expected values of both statistics are equal to the population mean, and since the standard deviation of the sample mean is smaller than that of the sample midrange, this tells us that, on average, the deviation of the sample mean from the population mean will be less than the deviation of the sample midrange from the population mean. Therefore the sample mean is preferable to the sample midrange as an estimator of the population mean.

Question Five

Hypotheses:

Let p_M and p_F be the proportions of all male and female parakeets (respectively) that can mimic speech.

$H_0: p_M = p_F$

$H_a: p_M \neq p_F$

Check of conditions and naming of test:

This is a two-proportion z-test.

We are told that we can consider the sets of male and female parakeets as random samples from the populations of male and female parakeets.

$n_M\hat{p}_M = 33 \cdot \frac{14}{33} = 14 > 10.$ $n_M(1 - \hat{p}_M) = 33 \cdot \frac{19}{33} = 19 > 10.$

$n_F\hat{p}_F = 44 \cdot \frac{13}{44} = 13 > 10.$ $n_F(1 - \hat{p}_F) = 44 \cdot \frac{31}{44} = 31 > 10.$

[Alternatively, one could check that $n_M\hat{p}_c$, $n_M(1 - \hat{p}_c)$, $n_F\hat{p}_c$, and $n_F(1 - \hat{p}_c)$ are all greater than 10, where $\hat{p}_c$ is the proportion of successful parakeets for the combined samples. Here $\hat{p}_c = (14 + 13)/(33 + 44) = 27/77$, making

$n_M\hat{p}_c = 33(27/77) = 11.57$, $n_M(1 - \hat{p}_c) = 33(50/77) = 21.43$,

$n_F\hat{p}_c = 44(27/77) = 15.43$, and $n_F(1 - \hat{p}_c) = 44(50/77) = 28.57$, which are all greater than 10.]

Mechanics:

$z = \dfrac{(\hat{p}_M - \hat{p}_F) - (p_M - p_F)}{\sqrt{\hat{p}_c(1 - \hat{p}_c)\left(\frac{1}{n_M} + \frac{1}{n_F}\right)}}$, where $\hat{p}_c$ represents the proportion of successful parakeets for the combined samples.

As explained above, $\hat{p}_c = 27/77$. So $z = \dfrac{(14/33 - 13/44) - (0)}{\sqrt{\left(\frac{27}{77}\right)\left(\frac{50}{77}\right)\left(\frac{1}{33} + \frac{1}{44}\right)}} = 1.172.$

So the p-value is 0.241. (This is calculated using 2*normalcdf(1.172,999), or the equivalent; on the TI-83 and 84 calculators, "normalcdf(" is accessed by selecting 2nd,DISTR,2.)

Conclusion:

Since the p-value is 0. 241, which is greater than 0.05, we do not have sufficient evidence to conclude that there is a difference between male and female parakeets in their ability to mimic speech.

Section II
Part B

Question Six

(a)

Student	1	2	3	4	5	6	7	8	9	10	11	12
Teacher X	99	94	88	80	83	84	82	82	82	91	90	65
Teacher Y	97	87	87	76	88	87	80	83	71	84	97	65
$d = X - Y$	2	7	1	4	−5	−3	2	−1	11	7	−7	0

Hypotheses:

Let μ_d be the mean difference in grades given by the two teachers.

$H_0: \mu_d = 0$

$H_a: \mu_d > 0$

[*Note: If you defined d as (Score from Teacher Y − Score from Teacher X), then your alternative hypothesis should be* $H_a: \mu_d < 0$.]

Check of conditions and naming of test:

This is a paired *t*-test.

We need to assume that the students given are a random sample of all possible students taught this course by these two teachers.

Since this is a small sample we consider a boxplot of the *d* values:

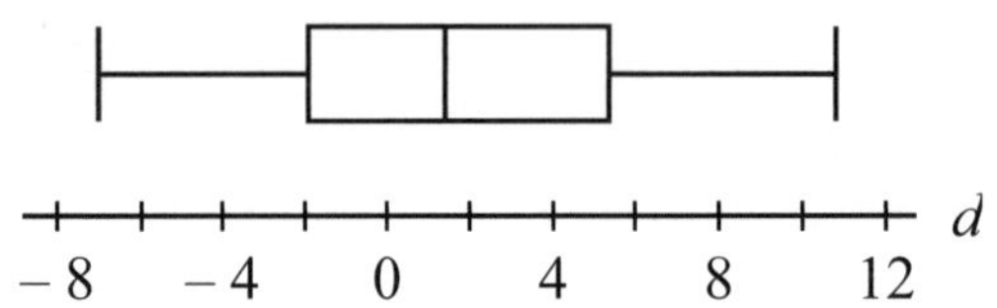

Since the distribution of the differences is roughly symmetrical and has no outliers, it appears reasonable to assume that the population of *d*-values is normally distributed. Therefore we can proceed with the paired *t*-test.

Mechanics:

$$t = \frac{\bar{d} - \mu_d}{\left(\frac{s_d}{\sqrt{n}}\right)} = \frac{1.5 - 0}{\left(\frac{5.231}{\sqrt{12}}\right)} = 0.993.$$

There are $n - 1 = 11$ degrees of freedom. So the *p*-value is 0.171 (This is calculated using tcdf(.993,999,11), or the equivalent; on the TI-83 and 84 calculators, "tcdf(" is accessed by selecting 2nd,DISTR,5.).

Conclusion:

Since the *p*-value is 0.171, which is greater than 0.05, we do not have sufficient evidence to conclude that the mean grade given by Teacher X is higher than the mean grade given by Teacher Y.

(b) Since the p-value is 0.015, which is less than 0.05, we have sufficient evidence to conclude that the mean grade given by Teacher A is higher than the mean grade given by Teacher B.

(c) No, the answer to part (b) does not imply that Teacher A is more generous than Teacher B. Since the students are assigned to the teachers according to what other courses they are taking, it is quite possible that the students taught by Teacher A are smarter on average than those taught by Teacher B, and so the fact that Teacher A is giving higher grades than Teacher B does not mean that Teacher A is more generous than Teacher B.

(d) The majority of the points in the scatterplot for Teacher A lie above the line $y = x$, which means that for most students, Teacher A is giving grades that are higher than those given by other teachers. In the scatterplot for Teacher B, the points are roughly evenly distributed above and below the line $y = x$, so generally speaking Teacher B is giving grades that are roughly the same as those given by other teachers. Therefore, in this sense, Teacher A appears to be more generous than Teacher B.

Sample Examination Five – Answers

1. **(B)** The 70th percentile weight is the weight below which 70% of all package weights lie. Now, 70% of 14,532 is $14532 \cdot (0.7) = 10172.4$. The number of packages with a weight less than 10 pounds is $3550 + 4215 = 7765$, and the number of packages with a weight less than 20 pounds is $3550 + 4215 + 2662 = 10427$. Therefore, the 70th percentile package weight lies between 10 and 20 pounds. (B) is the only response that satisfies this condition.

2. **(D)** The least squares regression line is defined to be the line that minimizes the sum of the squares of the vertical distances of the points from the line.

3. **(A)** For any number produced by the random number generator, the probability that the number is less than 3 is 2/5. Considering a number less than 3 as "success," we are interested in the number of successes in 10 trials, where the probability of success is the same for each trial and the outcomes of the trials are independent. Thus the number of successes is binomially distributed with $n = 10$ and $p = 2/5$, and so the required probability is the one given in (A).

4. **(C)** By emphasizing the cost of the renovations, the manager's question encourages the response "No." Thus, (C) is the correct response.

   Option (A) is incorrect since, although 12 members of the club failed to respond, these 12 members form only a small proportion of the club's membership, and therefore this is unlikely to be a source of bias. Option (B) is incorrect since the population of interest is the membership of the club, and all the members of the club are included in the survey. Options (D) and (E) are incorrect: the study does indeed make use of a census, but this is preferable to any form of sampling.

5. **(D)** Let X be the number of students who attend. We need

$$P(X \leq 4 \mid X \geq 2) = \frac{P(X \leq 4 \cap X \geq 2)}{P(X \geq 2)} = \frac{0.16 + 0.18 + 0.22}{0.16 + 0.18 + 0.22 + 0.16 + 0.08} = 0.7.$$

6. **(C)** By assigning each subject a random number between 0001 and 9999, ignoring repeats, and assigning the subjects with the lowest 50 numbers to Group 1, a completely randomized design is achieved. Note that in (B) the assignment is not completely random. Consider, for example, the last two subjects in the alphabetical list. These two subjects are very likely to be assigned to the same group. This cannot be the case in a completely randomized design. Options (D) and (E) are incorrect since the processes given don't involve any randomization. Option (A) is incorrect since the method described is unlikely to result in groups of equal size.

7. **(C)** Consider the third quartile, Q_3. Three quarters of the observations are smaller than Q_3 and one quarter are larger. Thus Q_3 is a measure of location and not a measure of spread. (The third quartile is not to be confused with the interquartile range (third quartile − first quartile), which *is* a measure of spread.)

8. **(B)** In the chi-square goodness of fit test the number of degrees of freedom is given by $k - 1$, where k is the number of outcome categories. Therefore, we use χ^2cdf(4.19,999,2), or the equivalent, which gives a p-value of 0.123. (On the TI-83 and 84 calculators, "χ^2cdf(" is accessed by selecting 2nd, DISTR, 7.) Since this value is greater than 0.05, we do not have sufficient evidence (at the 5% level) to conclude that the magazine's claim is incorrect. Note that (C) is incorrect, since we can never have convincing evidence that the null hypothesis is true. This would be "accepting H_0," which is never allowed.

9. **(B)** This is a two-proportion z-test. In significance tests for proportions, there can be no assumption of normality in a population, since our interest in the population is only with regard to a yes/no (success/failure) scenario. These tests use the normal distribution since, under the condition that the counts are large enough (as stated, for example, in response (A)), the distributions of the *sample proportions* are normal.

10. **(A)** The "R-sq" value given in the computer output is 22.1%. Thus, $r = \pm\sqrt{0.221} = \pm\, 0.470$. We know that the correlation between the total rainfall and the total sunshine is negative, since the output gives a negative slope (-1.670) for the regression line. Therefore, $r = -0.470$.

11. **(E)** The statistic referred to as "S" in the computer output is the standard error about the least-squares regression line, given by

$$s = \sqrt{\frac{\sum(y - \hat{y})^2}{n - 2}}.$$

The quantity $\sum(y - \hat{y})^2$ is the sum of the squares of the residuals, which are the vertical displacements of the points from the regression line. Thus s is a measure of the variability of the y-values from those predicted by the regression line.

12. **(B)** This is the significance test for the slope of the regression line, and the slope of the regression line for this sample is the coefficient of Rainfall in the regression equation. The p-value for the test is given in the "P" column of the computer output, in the "Rainfall" row, and the value given in the computer output in the context of this significance test is always the one for the two tailed version of the test. Thus the required p-value is 0.0178.

13. **(E)**

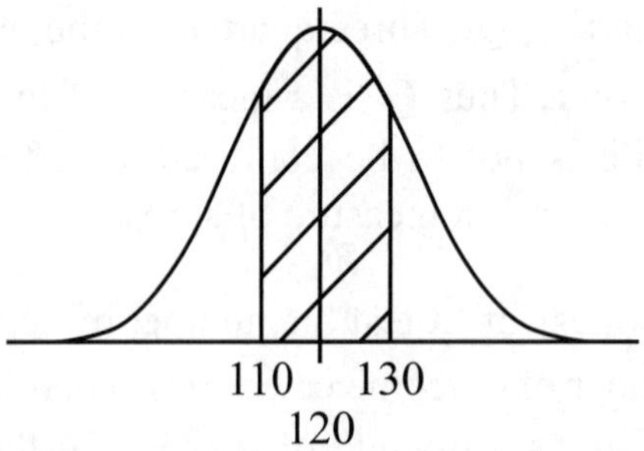

Since the mean of the given distribution is 120, we can see (as shown above) that 110 and 130 are equal distances to the left and the right, respectively, of the mean. The required probability is given by the area under the curve between 110 and 130, and it is easy to see that this is twice the area between 120 and 130.

Option (B) is incorrect, since the events "$X \geq 110$" and "$X \leq 130$" are not independent. (Knowing that $X \geq 110$ makes it less likely that $X \leq 130$.) Option (C) would be correct if the inequality in the subtracted probability were "*less* than or equal to," but this is not the case!

14. **(D)** When a population whose mean is μ and whose standard deviation is σ is normally distributed, it follows that the distribution of $(\overline{X} - \mu)/(\sigma/\sqrt{n})$ is normal (where $\overline{X}$ is the mean of a random sample and n is the sample size). However, when σ is unknown, the quantity $(\overline{X} - \mu)/(\sigma/\sqrt{n})$ cannot be used in the significance test, and so the quantity $(\overline{X} - \mu)/(s/\sqrt{n})$ is used in its place, and this quantity is t-distributed (with $n - 1$ degrees of freedom) if we know that the population is normally distributed. Thus (D) is the correct response. Note that (E) is incorrect since, as we know that the population is normally distributed, the sample mean, $\overline{X}$, is normally distributed.

15. **(D)** Let P(King) be p. Then P(Steward) $= p$, also. Moreover, P(Peasant) $= P$(Thief) $= 4p$. Since "King," "Steward," "Peasant," and "Thief" are the only four possibilities, $p + p + 4p + 4p = 1$. Therefore $10p = 1$, and so $p = 0.1$. Thus P(Peasant) $= 4(0.1) = 0.4$.

16. **(E)** This is a paired t confidence interval. We use a t-distribution with $n - 1 = 10 - 1 = 9$ degrees of freedom. Obtained from the table of t-distribution critical values, the critical value of this distribution for a 95% confidence interval is 2.262. So the confidence interval is given by $\bar{d} \pm t^* \cdot \frac{s_d}{\sqrt{n}} = 0.48 \pm 2.262 \cdot \frac{0.766}{\sqrt{10}}$.

17. **(C)** We use a z confidence interval for $\mu_2 - \mu_1$ since the population standard deviations are known. Thus the required confidence interval is given by $(\bar{x}_2 - \bar{x}_1) \pm z^* \sqrt{\frac{\sigma_1^2}{n_1} + \frac{\sigma_2^2}{n_2}}$. The z^* value, given by invNorm(.975), or the equivalent, is 1.96, and the values of $\sigma_1, \sigma_2, n_1, n_2$ are given in the question. (On the TI-83 and 84 calculators, "invNorm(" is accessed by selecting 2nd, DISTR, 3.) So all that remains for us is to calculate the value of $\bar{x}_2 - \bar{x}_1$. Now $\bar{x}_1$ will be the central value in the confidence interval for μ_1, and $\bar{x}_2$ will be the central value in the confidence interval for μ_2. Therefore $\bar{x}_1 = (14.223 + 16.164)/2 = 15.1935$, and $\bar{x}_2 = (16.388 + 18.930)/2 = 17.659$. So the required confidence interval is $(17.659 - 15.1935) \pm 1.96\sqrt{\frac{2.1^2}{18} + \frac{2.9^2}{20}} = (0.867, 4.064)$.

18. **(E)** Correlation does not imply causation: the fact that TV-watching has been found to be negatively correlated with test scores does not imply that watching a large amount of TV has *caused* low test scores. Therefore (E) is the correct response.

    The statement in (A) is correct (and therefore option (A) is incorrect) because the correlation of -0.898 does indicate a strong linear relationship between the variables. The statement in (B) is correct (and therefore option (B) is incorrect) since r^2 here is $(-0.898)^2 = 0.806$; thus, 80.6% (that is, *most*) of the variability in test scores can be explained by the regression line. The statements in (C) and (D) are correct (and therefore these options are incorrect) since the negative correlation indicates that a large amount of TV-watching was associated with low test scores, and vice versa.

19. **(C)** If the members of the treatment group were to undergo, on average, a significantly smaller amount of burning than the members of the control group, this would provide convincing evidence that the combination of the chemical and the liquid base (along with any possible placebo effect) provides protection against the sun. We will now think about why (A) and (B) are incorrect. The essence of experimental design is to obtain two groups that differ *only* with respect to the factor that is to be tested. So, in the context of this experiment, if the control group were to receive the liquid base (and no chemical), and if we were to make sure that that the complete lotion and the liquid-base-only were identical in color, smell, and so on, then the only difference between the two groups would be the presence of the chemical in the lotion given to the treatment group. Thus, with that design, the effectiveness of the chemical itself could be tested. However, with the experiment as it is described in the question, the groups differ with respect to the liquid base, the chemical, *and* the placebo effect (if you believe the placebo effect could happen in this context), and so it is impossible to distinguish between the effects of these three factors.

20. **(D)** In a significance test, the p-value is the probability, given that H_0 is true, that the observed outcome would take a value at least as extreme as the one actually observed.

21. **(E)** If we were given that the distribution of the taxable values of the homes in the city were normally distributed, then we would be able to determine the percentage of homes that were within two standard deviations of the mean. However, we are not given here that the distribution is normal. In fact, since we are told that the distribution is positively skewed, we know that the distribution is *not* normal. Therefore the percentage cannot be estimated from the information given.

22. **(B)** The probability of a Type I error is always equal to the significance level of the test (which is here given to be 0.05). So an increase in the sample size will have no effect on the probability of a Type I error. Therefore (B) is the correct response.

    An increase in the sample size *will*, however, decrease the probability of a Type II error; thus, (C) is an incorrect response. What is more, power is the probability that a Type II error is *not* made, and therefore an increase in the sample size will result in an increase in the power; thus, (D) is an incorrect response. Turning our attention to (E), the standard deviation of the sampling distribution of the sample proportion is $\sqrt{p(1-p)/n}$, where n is the sample size, and therefore this standard deviation is reduced by an increase in n; thus (E) is an incorrect response. Finally, considering (A), the confidence interval for the population proportion is $\hat{p} \pm z^* \sqrt{\hat{p}(1-\hat{p})/n}$. When the sample size, n, is increased, the sample proportion, $\hat{p}$, is unlikely to be changed greatly (what is more, any increase in the value of $\hat{p}$ will be compensated by a decrease in the value of $(1-\hat{p})$, and vice versa) and so the increase in the value of n in the denominator of the fraction is likely to result in a decrease in the width of the confidence interval; thus, response (A) is incorrect.

23. **(B)** Consider the sequence of t-distributions, as we increase the number of degrees of freedom. With low numbers of degrees of freedom, the t-distributions are noticeably different from the standard normal distribution and have greater variability than the standard normal distribution. As the number of degrees of freedom increases, the variability of the t-distributions decreases, and the t-distributions get closer to the standard normal distribution. Therefore, of the distributions given, the one with the smallest standard deviation is $N(0,1)$, followed by t_{10}, and then by t_5. (Please see "Understand where the t-distributions come from" in the Top Tips section of the Question Book for a graph that compares two of the t-distributions with the standard normal distribution.)

24. **(B)** Cluster sampling is carried out as follows. The population is divided into clusters, some of the clusters are selected at random, and then individuals from the selected clusters form the sample. Under the strategy given in (B), the resulting sample is likely to be representative of the population of third graders in terms of reading ability. In (C) and (D), certain parts of the population will necessarily be excluded from the sample, and so these options are not appropriate. To use the strategy given in (E) would defeat the object of cluster sampling, which is to cut down the work and expense involved: if there are many (small) clusters, then a large number of clusters would have to be selected in order to adequately represent the population, which would be time-consuming and expensive. Likewise, (A) is incorrect: the point of a cluster sample is to avoid the work and expense involved in sampling methods such as a simple random sampling.

25. **(D)** We can show that A and B are not independent a number of different ways: by showing that $P(A \cap B) \neq P(A) \cdot P(B)$, or that $P(A \mid B) \neq P(A)$, or that $P(B \mid A) \neq P(B)$, or by many other similar methods. We will use the first one.

    The total number of cars sold is $977 + 421 + 244 + 361 + 610 + 326 = 2939$.

    So $P(A) = (977 + 421)/2939 = 0.476$, $P(B) = (977 + 244 + 610)/2939 = 0.623$, and $P(A \cap B) = 977/2939 = 0.332$.

    Therefore $P(A) \cdot P(B) = (0.476)(0.623) = 0.297$, and this is not equal to $P(A \cap B)$. So the events A and B are not independent.

    To show that events A and B are not mutually exclusive, we have to show that (when one vehicle is picked at random) both events can happen, in other words that the vehicle can be both a sedan and bought from the downtown office. This clearly is possible, and so the events A and B are not mutually exclusive.

26. **(A)** The power of the test is the probability, when H_a is true, that H_0 is correctly rejected in favor of H_a.

27. **(C)** Let the proportion of all male students who approve be p_1 and the proportion of all female students who approve be p_2. When a random sample of 80 male students is taken, the proportion, $\hat{p}_1$, of the sample who approve has standard deviation $\sqrt{p_1(1-p_1)/80}$. Similarly, the proportion, $\hat{p}_2$, of the sample of 100 female students who approve has standard deviation $\sqrt{p_2(1-p_2)/100}$, and these two proportions are independent. Therefore, using the formula $\sigma_{X+Y} = \sqrt{\sigma_X^2 + \sigma_Y^2}$, we see that the standard deviation of $\hat{p}_1 + \hat{p}_2$ is $\sqrt{p_1(1-p_1)/80 + p_2(1-p_2)/100}$. To obtain the *mean* of the two sample proportions we divide $(\hat{p}_1 + \hat{p}_2)$ by 2, and so the standard deviation is divided by 2, also. Hence the standard deviation of $(\hat{p}_1 + \hat{p}_2)/2$ is $(1/2)\sqrt{p_1(1-p_1)/80 + p_2(1-p_2)/100}$.

Now the question requires the standard *error* of the mean of the two sample proportions. The standard error of a statistic is obtained by replacing, in the formula for the standard deviation of the statistic, all population parameters by their estimates obtained from the sample(s). Therefore, here, p_1 is replaced by $23/80$ and p_2 is replaced by $38/100$. This gives the required standard error to be

$$\frac{1}{2}\sqrt{\frac{(23/80)(57/80)}{80}+\frac{(38/100)(62/100)}{100}}.$$

28. **(E)** Within each block, we need the plots to be similar in terms of the amounts of water and light received. Looking at (E), we see that this is the case with all six blocks listed: plots 1 and 4 receive similar amounts of light and similar amounts of water, plots 2 and 3 receive similar amounts of light and similar amounts of water, and so on. Thus option (E) gives a suitable blocking scheme. All the other options give schemes that are unsuitable. For example, the first block in (A) contains plots 1 and 9, which receive different amounts of light; the first block in (B) contains plots 1 and 2, which receive different amounts of water; likewise, every block given in (A)-(D) contains plots that receive differing amounts of light and/or water. Thus (E) is the only option that provides a suitable blocking scheme.

29. **(D)** A (two-sided) 95% confidence interval is equivalent to a two-tailed test with a 5% significance level. So here, since 158.4 is contained in the confidence interval, we do not have convincing evidence at the 5% level that $\mu \neq 158.4$.

    Note that (A) is incorrect because we can never obtain convincing evidence that μ is *exactly* 158.4: this would be accepting H_0, which is never allowed. Options (B) and (C) are clearly incorrect since the confidence interval for μ contains 158.4.

    For anyone who is concerned that the statement in (E) might be true, consider the following (somewhat tricky) argument. In terms of critical values and tail probabilities, a one-tailed test at the 5% significance level is equivalent to a two-tailed test at the 10% significance level. Furthermore, a two-tailed test at the 10% significance level is equivalent to a 90% (two-sided) confidence interval. So we will use the information given to calculate a 90% confidence interval for μ. The center of the confidence interval given is $(158.28 + 160.92)/2 = 159.60$, and the margin of error (half of the width of the confidence interval) is $160.92 - 159.60 = 1.32$. Now, this (given) confidence interval was calculated using the t-distribution with 19 degrees of freedom, of which the critical value for a 95% confidence level is 2.093. The critical value of that same distribution for a 90% confidence level is 1.729. So the margin of error of the 90% confidence interval will be $1.32(1.729/2.093) = 1.090$. Therefore, the 90% confidence interval for μ is $159.60 \pm 1.090 = (158.51, 160.69)$. 158.4 is not in this interval, and thus we have convincing evidence at the 10% level that $\mu \neq 158.4$. Equivalently, since every value in the 90% confidence interval is *greater* than 158.4, we have convincing evidence at the 5% level that $\mu > 158.4$. Hence the statement in (E) is false.

30. **(D)** Since the total number of umbrellas is 300, we can calculate the total number of black umbrellas by $300 - 35 - 79 = 186$. Using this result, we can calculate the number of black telescopic umbrellas by $186 - 59 = 127$. From this, we see that the total number of telescopic umbrellas is $127 + 47 + 13 = 187$. Hence the expected number of black telescopic umbrellas is $(186)(187)/300 = 115.94$.

31. **(A)** Recall the following three facts: First, for *any* two random variables X and Y, $E(X+Y) = E(X) + E(Y)$. Second, if X and Y are *independent*, then $\sigma_{X+Y} = \sqrt{\sigma_X^2 + \sigma_Y^2}$. Third, that if X and Y are *independent* and normally distributed then the random variable $X + Y$ is normally distributed. Since here it's a married couple being chosen at random (as opposed to a man being chosen at random and a woman being chosen at random), the height of the man and the height of the woman are not independent. Thus, only statement I can be concluded from the information given.

32. **(C)** The 95% confidence interval for μ is given by $\bar{x} \pm t^*(s/\sqrt{n})$. So the center of the confidence interval is at the sample mean, $\bar{x}$, and the distance from the center of the interval to the upper (or lower) end of the interval is given by $t^*(s/\sqrt{n})$. Only (C) and (E) give intervals for which the center is at the sample mean, and so (A), (B) and (D) are eliminated. Now t^*, the critical value of the t-distribution with 14 degrees of freedom, is 2.145, and $n = 15$. Looking at the dotplot given, we estimate the standard deviation of the sample (a typical deviation from the sample mean) to be somewhere between 15 and 30. (The true value of the sample standard deviation is close to 25.) So the distance from the center of the interval to the upper end of the interval is between $2.145(15/\sqrt{15}) = 8.3$ and $2.145(30/\sqrt{15}) = 16.6$. The interval given in (B) satisfies this condition, and the one given in (E) does not.

33. **(E)** Recall that if a and b are constants, x and y are defined as they are in the question, and $y = ax + b$, then $\bar{y} = a\bar{x} + b$ and, if $a > 0$, $s_y = as_x$. In other words, multiplying by a multiplies the mean and the standard deviation by a, and adding b adds b to the mean but has no effect on the standard deviation (and similarly subtracting b would subtract b from the mean and would have no effect on the standard deviation). Now consider (E). We start with a mean of 52.3 and a standard deviation of 10.8. Subtracting 52.3 will reduce the mean by 52.3 making the mean now 0 and leaving the standard deviation at 10.8. We then multiply by 7.9/10.8, which will transform the standard deviation to 7.9 and will leave the mean at 0. Finally, adding 80.6 will increase the mean to 80.6 and will leave the standard deviation at 7.9. These are the required mean and standard deviation.

34. **(D)** Suppose, for example, the sample size used was 100. A chi-square test for goodness of fit can be used, setting up a table as shown below.

	Observed Count	Expected Count
Males	*	50
Females	*	50

The chi-square goodness of fit test can then be performed in the usual way. (For interest, it is worth noting that the p-value obtained for the chi-square goodness of fit test will be exactly the same as the p-value obtained for the one-proportion z-test.)

Note that (E) is incorrect since only one species is being considered. If two or more species were being compared in terms of the proportions of the species that are males, then the chi-square test for homogeneity could be used. Likewise, (C) is incorrect: there is only one sample in this study. Options (A) and (B) are incorrect since this study is concerned with the *proportion* of males in the species; thus, tests for means cannot be used to investigate the biologist's question.

35. **(D)** By definition, the sample of 81 employees will be a simple random sample if (and only if) every subset of 81 employees has an equal chance of being the sample. This is not the case here since, for example, a sample that includes 51 manual workers is not possible. Thus, not every subset of 81 employees has an equal chance of being the sample, and so the sample obtained will not be a simple random sample. (Considering option (A), we note that, since exactly 1/10 of each type of employee is included in the sample, every employee has an equal probability (1/10) of being included in the sample. However, this does *not* enable us to conclude that the sample obtained will be a simple random sample.)

36. **(D)** When a population with mean μ and standard deviation σ is normally distributed, and a random sample of size n is taken from the population, the distribution of the sample mean is $N(\mu, \sigma/\sqrt{n})$. So here the sample mean battery life has distribution $N(\mu, 4.3/\sqrt{10})$. The graph of this distribution is shown below.

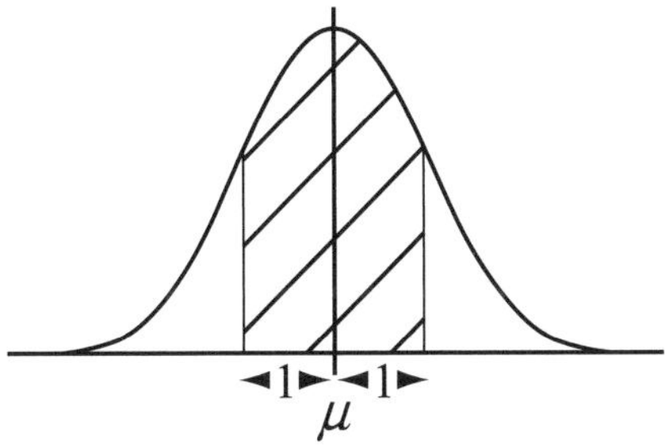

The required probability is the area of the shaded region. Since the right boundary of this region is 1 month above the mean, the z-value there is given by

$$z = \frac{1}{4.3/\sqrt{10}} = 0.7354.$$

Similarly, the z-value at the left boundary is -0.7354. Therefore the required probability is obtained using normalcdf(-.7354,.7354), or the equivalent, which gives 0.538. (On the TI-83 and 84 calculators, "normalcdf(" is accessed by selecting 2nd, DISTR, 2.)

37. **(A)** First look at the scatterplot shown above the question. Remember that a point will have a positive residual if it lies above the regression line, and a negative residual if it lies below the regression line. So if we look across the scatterplot from left to right, we see negative residuals, then primarily positive residuals, then primarily negative residuals. Looking at the residual plot in (A) we have exactly that: negative residuals, then primarily positive residuals, then primarily negative residuals. This is not the case in any of the other options.

38. **(C)** A sample proportion of $\hat{p} = 107/200 = 0.535$ has been obtained. The formula for the 95% confidence limits for p is $\hat{p} \pm z^* \sqrt{\hat{p}(1-\hat{p})/n}$. We use $\hat{p} = 0.535$, $z^* = 1.96$ (given by invNorm(.975), or the equivalent), and $n = 200$. This gives a confidence interval of (0.466, 0.604). (On the TI-83 and 84 calculators, "invNorm(" is accessed by selecting 2nd, DISTR, 3.) In other words, we are 95% confident that p is between 0.466 and 0.604, as stated in (C).

Let's be clear that (D) is incorrect. The equivalent true statement would be that if the procedure of flipping the coin 200 times were to be repeated a large number of times, then 95% of the resulting confidence intervals would contain the true proportion of "heads" for this coin. Both this correct statement and the correct statement in (C) tell us something about the *population* proportion. The statement in (D) tries to relate the confidence interval to *sample* proportions. Likewise, the statement in (E) tries to relate the confidence interval to a sample proportion, and therefore is incorrect.

39. **(C)** In any cumulative relative frequency graph, such as the one given in (C) for example, the cumulative relative frequency plotted for a measurement of 6, say, is the proportion of all measurements that are less than or equal to 6. Likewise, the cumulative relative frequency plotted for a measurement of 7 is the proportion of all measurements that are less than or equal to 7. So for this data set, the difference between these two cumulative relative frequencies must be the proportion of all measurements that are equal to 7, which is governed by the height of the block in the histogram above 7 on the horizontal axis. Therefore the amounts by which the cumulative relative frequencies increase (the vertical increments in the cumulative relative frequency graph) are governed by the heights of the blocks in the histogram. The heights of the blocks in the histogram increase up to 8 on the horizontal axis, and therefore we are looking for a cumulative relative frequency graph for which the vertical increments are increasing up to 8 on the horizontal axis. This is the case only for (C).

40. **(E)** We will first calculate the probability of getting red, green, blue, in that order. There are 10 beads in total and five of them are red, and so the probability that the first bead *is* red is 5/10. If the first bead is red, then there are 9 beads remaining and 3 of them are green. So the probability that the next bead is green is 3/9. Once this has happened, there are 8 beads remaining and 2 of them are blue. So the probability that the last bead is blue is 2/8. Therefore the probability of getting red, green, blue, in that order is given by

$$P(RGB) = \left(\frac{5}{10}\right)\left(\frac{3}{9}\right)\left(\frac{2}{8}\right).$$

Now we'll consider a different order of those three colors: RBG. By the same logic, the probability of this is given by

$$P(RGB) = \left(\frac{5}{10}\right)\left(\frac{2}{9}\right)\left(\frac{3}{8}\right),$$

which has the same value as that of $P(\text{RGB})$. It should be clear now that every different order of the three colors has the same probability.

So all that remains is to calculate the number of possible orders of the three colors. The orders can be listed as follows: RGB, RBG, GRB, GBR, BRG, BGR. So there are 6 possible orders. Therefore we need 6 times the probability calculated above, which is the quantity given in (E).

Section II

Question One

(a)

Number of Breakdowns	Frequency 1952-1978	Frequency 1979-2008
0	2	2
1	3	7
2	7	11
3	6	7
4	4	1
5	4	1
6	1	1

(b) A back-to-back histogram displaying these data is shown below.

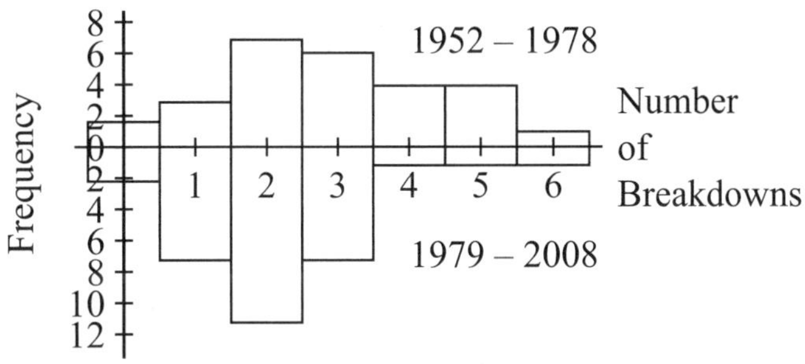

[*Several other types of display are possible here; for example, histograms drawn side-by-side and with the same scales, a bar graph with columns in pairs (shaded differently to indicate the two distributions), parallel boxplots with a single scale, or frequency polygons drawn on a single pair of axes.*]

(c) Based on the display in part (b), the median of the distribution for 1979-2008 appears to be around 2 breakdowns per year, and this is less than the median of the distribution for 1952-1978, which appears to be around 3 breakdowns per year. Therefore the occurrence of breakdowns does seem to have decreased since the time of the refurbishment.

(d) The distribution of the annual numbers of breakdowns in the years 1952 to 1978 is roughly symmetrical. The distribution of thc annual numbers of breakdowns in the years 1979 to 2008 is positively skewed (skewed to the right).

[*The histogram for 1952-1978 could be said to give an impression of positive skewness, and therefore a statement that this distribution is positively skewed would be accepted. Inspection of the boxplot for this distribution shows that the distribution, in terms of the locations of its maximum, minimum, median, and quartiles, is very close to being symmetrical.*]

Question Two

(a) Number the students in the school from 1 to N, where N is the total number of students. Use a random number generator to select an integer between 1 and N, inclusive. The student with this number will be the first student in the sample. Repeat this process, ignoring repeated numbers, until 40 students have been selected.

(b) Using a method similar to the one used in (a), randomly select 10 students from the 9th grade. In the same way, randomly select 10 students from the 10th grade, 10 students from the 11th grade, and 10 students from the 12th grade. These 40 students will form the sample.

(c) It is sensible to stratify by grade level if we believe that the four grades might differ from one another in terms of the amount of time spent studying and that within each grade the amount of time spent studying varies little between students. For example, the statistics students might well suspect, in advance of doing the survey, that older students tend to study longer than younger students, with some degree of consistency of study time within the grades.

(d) Assume that the criteria mentioned in part (c) are satisfied. Then, using a stratified sample in the way described in place of a simple random sample will reduce the variability of the estimate of the population mean, in the sense that if many of these stratified samples were taken, their means would vary less than the means of a large set of simple random samples. This would come about as a result of the fact that each stratified sample will have 10 students from each grade, whereas the simple random samples would have varying numbers of students from the four grades. Thus stratified sampling will produce a more reliable estimate of the population mean.

Question Three

(a) At Machine A, the probability of a positive gain is
$P(X > 0) = 0.20 + 0.13 + 0.07 = 0.4.$

At Machine B, the probability of a positive gain is
$P(Y > 0) = 0.16 + 0.10 + 0.05 = 0.31.$

(b) The expected value of the money gained in any play at Machine A is
$\mu_X = (-5)(0.33) + (-2)(0.27) + (1)(0.20) + (4)(0.13) + (10)(0.07) = -\$0.77.$

The standard deviation of the money gained in any play at Machine A is

$$\sigma_X = \sqrt{(-5-(-0.77))^2(0.33) + (-2-(-0.77))^2(0.27) + (1-(-0.77))^2(0.20) + (4-(-0.77))^2(0.13) + (10-(-0.77))^2(0.07)} = \$4.24$$

The expected value of the money gained in any play at Machine B is
$\mu_Y = (-5)(0.37) + (-2)(0.32) + (1)(0.16) + (4)(0.10) + (25)(0.05) = -\$0.68$

The standard deviation of the money gained in any play at Machine B is

$$\sigma_Y = \sqrt{(-5-(-0.68))^2(0.37) + (-2-(-0.68))^2(0.32) + (1-(-0.68))^2(0.16) + (4-(-0.68))^2(0.10) + (25-(-0.68))^2(0.05)} = \$6.56$$

(c) No, the friend's argument in not valid, since the *amount* of the gain or loss is relevant in the choice of machine, rather than just whether the gain is positive or negative. If a large number of plays is made at Machine A, we expect the mean gain per play to be around the expected value for Machine A, which is $-\$0.77$. Likewise, if a large number of plays is made at Machine B we expect the mean gain per play to be around $-\$0.68$. Therefore, Machine B is the better choice.

[*It is worth noting here that due to the large standard deviations (compared to the sizes of the means) it is still quite possible, in a large number of plays, that Machine A could turn out to be the better choice. For example, suppose that 100 plays were to be made at Machine A and 100 plays were to be made at Machine B. Using the Central Limit Theorem we can show that the probability of the mean gain at Machine A being greater (less negative) than the mean gain at Machine B is 0.45. Nonetheless, this confirms that more often than not, when a large number of plays is made, Machine B is the better choice.*]

Question Four

Hypotheses:

H_0: The distributions of car type are the same in the populations of reservations at all three locations.

H_a: The distributions of car type are not all the same.

Check of conditions and naming of type of test:

This is a chi-square test for homogeneity.

The conditions for the chi-square test for homogeneity are:

1. Random samples
2. Large sample sizes

We are told that we can consider the samples of reservations as random samples from the populations of reservations at the three locations. So condition 1 is satisfied.

The expected counts are shown in the table below.

	Type			
	Compact	Standard	Full Size	Other
Location 1	81.98	67.9	46.31	53.82
Location 2	90.83	75.23	51.31	59.63
Location 3	89.19	73.87	50.38	58.55

Since all the expected counts are greater than 5, the sample sizes are large enough for the chi-square test to be appropriate. Thus condition 2 is satisfied.

Mechanics:

$$\chi^2 = \sum \frac{(\text{Obs} - \text{Exp})^2}{\text{Exp}}$$

$$= 0.436 + 0.743 + 0.859 + 0.499 + 0.008 + 0.102 + 0.362 + 0.094 + 0.520 + 1.319 + 2.237 + 0.974$$

$$= 8.154$$

We use the chi-square distribution with $(3 - 1)(4 - 1) = 6$ degrees of freedom.

So the p-value is $P(\chi^2_6 > 8.154) = 0.227$. (This result is obtained using χ^2cdf(8.154,999,6), or the equivalent; on the TI-83 and 84 calculators, "χ^2cdf(" is accessed by selecting 2nd,DISTR,7.)

Conclusion:

Since the p-value is 0.227, which is greater than 0.05, we do not have statistically significant evidence of a difference in the distributions of the car types between the populations of reservations at the three locations.

Question Five

(a) The scatterplot, with the least squares regression line added, is shown below.

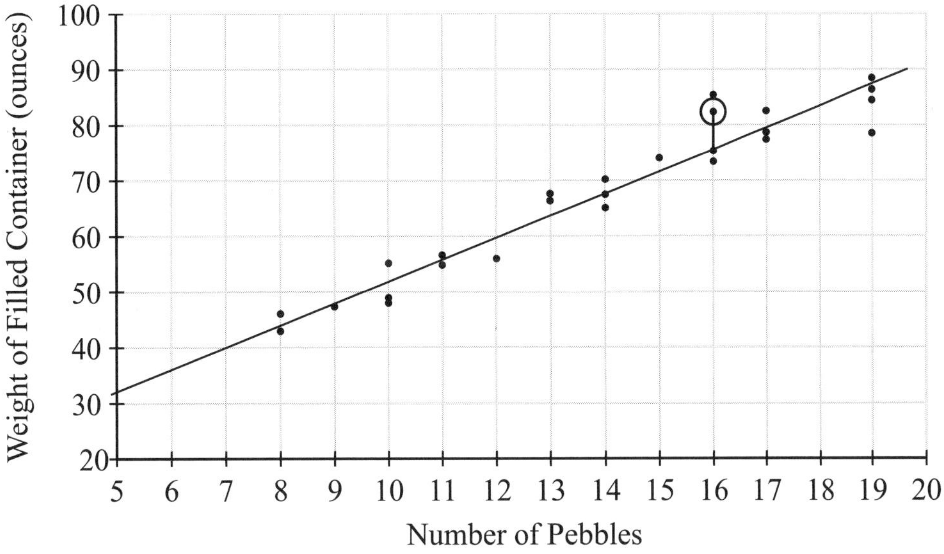

[*The easiest way to draw the line is to plot the points for* $x = 5$ *and* $x = 20$ *first, and then to join them with a straight line. (Be sure to use a straight edge for drawing the line!) The y-coordinate of the point with x-coordinate* 5 *is found by substituting* $x = 5$ *into the equation of the regression line, giving* $\hat{y} = 13.419 + 3.865 \cdot (5) = 32.7$. *Similarly, the y-coordinate of the point with x-coordinate* 20 *is* $\hat{y} = 13.419 + 3.865 \cdot (20) = 90.7$.]

(b) The point (16, 82) has been circled and the line segment corresponding to the residual for this point has been drawn on the scatterplot given in the solution to part (a).

The predicted value of y for $x = 16$ is given by $\hat{y} = 13.419 + 3.865 \cdot (16) = 75.26$. So the residual for the point (16, 82) is $82 - 75.26 = \boxed{6.74 \text{ ounces}}$.

(c) The weight of the filled container for this student was 6.74 ounces more than the weight predicted by the regression line for a student who collected 16 pebbles.

(d) The values of the slope and the intercept of the least squares regression line both have meaningful interpretations in this context. The slope, 3.865, is the number of ounces by which the predicted weight of the filled container increases for each increase of one in the number of pebbles collected. The intercept, 13.419, is the value predicted by the regression line for the weight (in ounces) of the "filled container" when the number of pebbles collected is zero. This value, 13.419 ounces, is therefore an estimate of the weight of the empty container.

[*Note that since the slope is an estimate of the extra weight of the filled container when the number of pebbles collected is increased by one, it would make sense for this value,* 3.865 *ounces, to be the mean weight of all the pebbles collected by the children. If we assume that the weight of the empty container is* 13.419 *(as estimated by the regression line), then this is indeed the case. Therefore, it is acceptable to use this alternative interpretation of the slope in part (d).*]

Section II
Part B

Question Six

(a) Hypotheses:
Let p be the proportion of the members of this gym who have exercise equipment at home.

$H_0: p = 0.6$

$H_a: p < 0.6$

Check of conditions and naming of test:
This is a one-proportion z-test.

[*Note that since the type of test is named in the question it is unnecessary to name it in your solution. Nonetheless we include it here for the sake of completeness.*]

The conditions for this test are:
1. Simple random sample
2. np_0 and $n(1-p_0)$ both greater than 10

It is stated in the question that we can assume that the 83 people in Amy's sample form a random sample of the gym's members. Moreover, $np_0 = 83 \cdot 0.6 = 49.8$ and $n(1-p_0) = 83 \cdot 0.4 = 33.2$, which are both greater than 10.

Mechanics:

$$z = \frac{\hat{p} - p_0}{\sqrt{\frac{p_0(1-p_0)}{n}}} = \frac{42/83 - 0.6}{\sqrt{\frac{(0.6)(0.4)}{83}}} = -1.748.$$

Therefore the p-value is $P(Z < -1.748) = 0.040$. (This result is obtained using normalcdf(−999,−1.748), or the equivalent; on the TI-83 and 84 calculators, "normalcdf(" is accessed by selecting 2nd, DISTR, 2.)

Conclusion:
Since the p-value is 0.040, which is less than 0.05, the result provides convincing evidence that fewer than 60% of the gym's members have exercise equipment at home.

(b) In the proposed one-proportion z-test, p_0 would be 0.6 and n would be 10, making $np_0 = 10 \cdot (0.6) = 6$ and $n(1-p_0) = 10 \cdot 0.4 = 4$, neither of which is greater than 10. Consequently, this one-sample z-test cannot be used.

(c) For each person in the sample, the probability of having exercise equipment at home is 0.6. Moreover, since we're considering a *random* sample of 10 of the gym's members, and since we're told that we can assume that the gym has a large number of members, the outcomes for the 10 people in the sample are independent. Therefore, denoting the number of people in the sample who have exercise equipment at home by X, the distribution of X is binomial, with $n = 10$ and $p = 0.6$.

Hence,

$$\begin{aligned} P(X \le 2) &= P(X = 0) + P(X = 1) + P(X = 2) \\ &= (0.4)^{10} + {}_{10}C_1(0.6)(0.4)^9 + {}_{10}C_2(0.6)^2(0.4)^8 \\ &= \boxed{0.012}. \end{aligned}$$

(On the TI-83 and 84 calculators, ${}_{10}C_2$, for example, is obtained by typing "10 nCr 2"; "nCr" is accessed by selecting MATH, PRB, 3.)

(d) Yes. The answer to part (c) tells us that if exactly 60% of the gym's members had exercise equipment at home then the probability of getting 2 or fewer in a random sample of 10 of the gym's members having exercise equipment at home is 0.012, which is small. In other words, if exactly 60% of the gym's members had exercise equipment at home, then a sample result as low as 2, as obtained in Leonard's sample, would be very unlikely. Therefore we reject the hypothesis that 60% of the gym's member's have exercise equipment at home in favor of an alternative hypothesis that fewer than 60% of the gym's member's have exercise equipment at home.

NOTES

NOTES

NOTES

NOTES

NOTES

NOTES